Writing Intimacy into Feminist Geography

Intimacy, expressed through the feelings and sensations of the researcher, is bound up in the work of a feminist geographer. Tapping into this intimacy and including it in academic writing facilitates a grasping of the effects of power in particular places and initiates a discussion about how to access and tease out what constitutes the intimate both ethically and politically throughout the research process.

This collection provides valuable reflections about intimacy in the research process - from encounters in the field, through data analysis, to the various pieces of written work. A global and heterogeneous pool of scholars and researchers introduce personal ways of writing intimacy into feminist geography. Authors expand existing conceptualizations of intimacy and include their own stories as their chapters explore the methodological challenges of using intimacy in research as an approach, a topic and a site of interaction.

The book is valuable reading for students and researchers of Geography, as well as anyone interested in the ethics and practicalities of feminist, critical and emotional research methodologies.

Pamela Moss is a Professor in Human and Social Development, University of Victoria, Canada.

Courtney Donovan is an Associate Professor in the Department of Geography and Environment, San Francisco State University, USA.

Writing Intimacy into Feminist Geography

**Edited by Pamela Moss and
Courtney Donovan**

Routledge
Taylor & Francis Group

LONDON AND NEW YORK

First published 2017
by Routledge

2 Park Square, Milton Park, Abingdon, Oxfordshire OX14 4RN
52 Vanderbilt Avenue, New York, NY 10017

Routledge is an imprint of the Taylor & Francis Group, an informa business

First issued in paperback 2018

British Library Cataloguing in Publication Data
A catalogue record for this book is available from the British Library

Library of Congress Cataloging in Publication Data
Names: Moss, Pamela, 1960- editor. | Donovan, Courtney, editor.
Title: Writing intimacy into feminist geography / edited by Pamela Moss and Courtney Donovan.
Description: Abingdon, Oxon ; New York, NY : Routledge, [2017] | Includes bibliographical references and index.
Identifiers: LCCN 2016042608| ISBN 9781472476777 (hbk) | ISBN 9781315546186 (ebk)
Subjects: LCSH: Feminist geography--Methodology. | Intimacy (Psychology)
Classification: LCC HQ1233 .W748 2017 | DDC 158.2—dc23
LC record available at https://lccn.loc.gov/2016042608

ISBN: 978-1-4724-7677-7 (hbk)
ISBN: 978-0-367-13878-3 (pbk)

Typeset in Times New Roman
by Swales & Willis Ltd, Exeter, Devon, UK

Contents

Figures

Contributors

Gail Adams-Hutcheson (BSocSc, MSocSc, PhD) is a Teaching Fellow at the University of Waikato, New Zealand. She teaches in the areas of gender, feminism, the body and emotional and affective geographies. Gail was an editorial assistant to Professor Lynda Johnston for *Gender, Place and Culture: A Journal of Feminist Geography*. Her recently completed doctoral thesis spans the emotional and affective lifeworld of people who relocated out of Christchurch after the earthquakes and aftershocks. She has published journal articles on the place of psychoanalysis in geographical methods ('Methodological reflections on transference and countertransference in geographical research', *Area*, 2013) and the postgraduate experience of changing academic landscapes ('Postgraduates performing powerfully in a changing academic environment', *New Zealand Geographer*, 2014).

Toni Alexander (BA, MA, PhD) is currently an Associate Professor and Chairperson of the Department of Global Cultures and Languages at Southeast Missouri State University, USA. Her training was in geography with an emphasis upon cultural geographies of socially and spatially marginalized minority groups. Her research efforts have used qualitative and archival approaches to understand the Chinese immigrant experience during the California Gold Rush and that of poor white domestic 'Okie' migrants to the state throughout the twentieth century. In addition, she has also explored the cultural impacts of contemporary Hispanic immigration to a rural southern Black Belt community. Her work in these various communities has resulted in peer-reviewed scholarly journal publications and numerous professional conference presentations on both the national and international scales. As part of a National Science Foundation funded project, her interests in socio-spatial marginalization within higher education were addressed through a study of female faculty retention in Science, Engineering and Mathematics (SEM) programmes at large four-year institutions across the United States. Her efforts in this large-scale interdisciplinary effort resulted in the publication of an invited, peer-reviewed chapter in *Mentoring Strategies to Facilitate the Advancement of Women Faculty*, published by the American Chemical Society, as well as a peer-reviewed article in the *International Journal of Diversity Organizations, Communities, and Nations*.

Kye Askins (BSc, PhD) is a Reader in Urban Geography at the University of Glasgow, Scotland. In her pre-academic life, she worked within the homelessness and mental health sectors, then studied Environmental Management and enjoyed a career in the community composting world. She came to Human Geography through a PhD at the University of Durham, which was the start of her ongoing research interests in identity, citizenship, emotions and everyday geographies of agency and resistance. She works from postcolonial and participatory paradigms, with a central aim of actively engaged research that, both theoretically and methodologically, challenges dominant discourses, foregrounds participants as co-producers of knowledge and works towards positive change with participants, local communities and students. She has written on issues of emotion, affect, materiality, feminist praxis and participatory geographies across a range of journals including *Emotion, Space and Society*, *Environment and Planning D*, *Area* and *Antipode*, as well as contributing chapters to *Placing Critical Geography*, *The Ashgate Research Companion to Critical Geopolitics* and *Fear: Critical Geopolitics and Everyday Life*.

Kathryn Besio (BA, MA, PhD) is an Associate Professor in the Geography and Environmental Studies Department at the University of Hawai'i at Hilo. She does research on a range of topics related to cultural and feminist geography, that include postcolonial tourism in Pakistan, feminist methodologies, dolphin watching and tourism in New Zealand and more recently issues related to food, bodies and cuisine in Hawai'i. Her work has been published in a number of geography journals, most notably *Gender, Place and Culture*, *The Professional Geographer*, *Environment and Planning A* and *Environment and Planning D*. She is currently working on a book length manuscript on home gardens in Hilo, Hawai'i that addresses the material and discursive significance of home garden spaces.

Dana Cuomo (BA, MA) is a PhD candidate in the departments of Geography and Women's Studies at Penn State University, USA. Her dissertation examines the policing and prosecution response to intimate partner violence to understand the links between gender and citizenship, the spatiality of law and policing, and state and intimate violence. Dana's research draws on her former employment experiences as a victim advocate serving survivors of domestic and sexual violence. She employs a feminist methodological research praxis to navigate conducting research with participants with whom she shares intimate, personal and ongoing relationships.

Sarah de Leeuw (BFA, MA, PhD), a creative writer and human geographer, is an Associate Professor in the University of Northern British Columbia's Northern Medical Program, in the Faculty of Medicine at the University of British Columbia, Canada. A two-time recipient of a CBC Literary Prize for Creative Non-Fiction and the holder of Northern British Columbia's first Michael Smith Foundation for Health Research Partnered Scholar Award, she teaches and undertakes research through critical and creative method/ologies about Indigenous and non-Indigenous relations in colonial landscapes, health and social inequalities and medical humanities. Author of three literary texts, including *Geographies of a Lover*, which, in 2013, was awarded the Dorothy Livesay BC Book Prize for the best book of poetry in British Columbia, Sarah has also co-edited two recent texts about intersectionality and determinants of Aboriginal peoples' well-being in Canada. Her poetry and creative non-fiction, some of which has been nominated for national and provincial literary prizes, appear in top-ranked literary journals, including *PRISM International*, *The Fiddlehead*, *ARC Poetry* and *CV2*. In addition to appearing in many handbooks and anthologies, her research has been widely published in academic journals ranging from *The Annals of the American Association of Geographers*, *cultural geographies*, *The Canadian Geographer*, *The International Journal of Mental Health and Addictions* and *The Journal of Canadian Family Physicians*. In 2015 and 2017 two new books of Sarah's (one a book-length long poem, the other a collection of creative non-fiction essays) will be released by Caitlin Press and NeWest Press, respectively. Sarah grew up on Haida Gwaii and now lives in Prince George.

Courtney Donovan (BA, MA, PhD) is Associate Professor of Geography at San Francisco State University, California. Trained as a geographer, her current research sits at the intersection of health geography, medical humanities, narrative medicine and the social determinants of health. She explores how art and the humanities can convey the personal narratives and health experiences of individuals and communities that experience the burden of marginalization and poor health outcomes. She conducts research in a number of sites, including graphic memoir, neighborhoods in transition and art therapy programs for soldiers. She has published in interdisciplinary journals as well as in the Companion to Health and Medical Geography. She is working on a number of projects that pull together her interests in health disparities, narrative medicine and digital health.

Karen Falconer Al-Hindi (BA, MA, PhD), **Pamela Moss** (BA, MA, PhD), **Leslie Kern** (BA, MA, PhD) and **Roberta Hawkins** (BA, PhD) began to explore collective biography together as a feminist research methodology in 2012. They began with the premise that academic work offers scholar-teachers more than neoliberalism promises: that it offers joy. Their investigation of joy in academia unfolded alongside a shared, expanding understanding of collective biography and a deepening intimacy. As their collaborative practices accumulated they recognized that knowing one another, gentle curiosity about each other's experiences, and intervention into one another's written memories, as well as a mutual authorial voice and use of the collaborative biography approach, reflected crosscutting layers of intimacy. Their previous work on this topic appears in *Social & Cultural Geography* (2014) and *Geography Compass* (2016).

Kathryn Gillespie (PhD) is a part-time lecturer at the University of Washington, USA in Geography, the Honors programme, and the Comparative History of

Ideas programme. Her work focuses on uneven relationships of power and privilege between human and nonhuman animals, with a particular emphasis on making legible the intimate lives of animals through narrative and multi-species ethnographic writing. She is currently working on a book based on her dissertation research, *The Cow with Ear Tag #1389*, about the lives of cows in the US dairy industry. She is co-editing two books, *Critical Animal Geographies: Politics, Intersections and Hierarchies in a Multispecies World* (Routledge, 2015) and *Economies of Death: Economic Logics of Killable Life and Grievable Death* (Routledge, forthcoming), and has published articles in *Gender, Place and Culture* and *The Brock Review.*

Kelsey B. Hanrahan (BSc, MA) is currently a doctoral candidate in the Department of Geography at the University of Kentucky in Lexington, USA. She holds a BSc in Archaeology (University of Calgary, Canada) and an MA in Anthropology (University of South Carolina, USA). Her research interests include livelihoods, care and social well-being, ageing, inter-generational rela-tions and personal relationships with a regional focus on West Africa. She also explores feminist methodologies including critical considerations of the fieldwork experiences and working with assistants and interpreters. She has presented at regional and national conferences and has published in *Gender, Place and Culture* and *ACME.*

Blake Hawkins (BA) is completing a Master of Library and Information Studies Degree at the University of British Columbia, in Vancouver BC, Canada. He is from the northwest coast of British Columbia, which has greatly influenced his research interests. Broadly, he has researched roles that gender, rural-ity and place can play in individual and community health, senses of place and well-being in Northern British Columbia. Such work has given him the opportunity to engage with feminist research methods (e.g. autobiography and autoethnography) for intimate examinations of these interests. To date, he has presented several papers on autoethnography, rural geography and indigenous/non-indigenous relations, at regional, national and international conferences. Currently, he has two related publications under review. While now situated in Vancouver, Blake returns to northern British Columbia several times a year.

Samuel Henkin (BA, MA) is currently a PhD student of Geography at the University of Kansas, USA. His broad research interests are in the spatialities of life and death, geographies of violence and the geographies of genocide. His research aims to expose and deconstruct violence that is both systemic in societal relations and embodied as an individual experience. His publications vary but all advance a greater understanding of everyday violence by linking critical theoretical approaches like feminist geopolitics and emotional geogra-phies and critical methodologies like discourse analysis. He has most recently participated as a research assistant for a National Science Foundation research project examining the spatialities of the Cambodian Genocide.

Robyn Longhurst (BSocSc, MSocSc, PhD) is Professor of Geography and Pro Vice-Chancellor (Education) at University of Waikato, New Zealand. She teaches in areas of gender, feminism, the body and social justice. Robyn has been Editor-in-Chief of *Gender, Place and Culture: A Journal of Feminist Geography* and Chair of the International Geographical Union Commission on Gender and Geography. She has published on issues relating to pregnancy, sexuality, mothering, 'visceral geographies', masculinities and body size and shape. Robyn is author of *Bodies: Exploring Fluid Boundaries* (Routledge, 2001), *Maternities: Gender, Bodies and Spaces* (Routledge, 2008) and co-author of *Pleasure Zones: Bodies, Cities, Spaces* (Syracuse University Press, 2001) and *Space, Place, and Sex: Geographies of Sexualities* (Rowman & Littlefield, 2010).

Clare Madge (BSc, PhD) is a Reader in the Department of Geography at the University of Leicester, UK. Her research interests are eclectic, ranging through postcolonial and feminist sensibilities, social media, and internet mediated research and creative forms of world-writing. At present, she is particularly interested in exploring the potentials of intimate poetry and autobiographical photography for expressing geographical worlds. She has published widely in books and journals both within and beyond Geography.

Vanessa A. Massaro (BA, MA) is a Visiting Assistant Professor in the Geography Department at Bucknell University, USA. She is finishing a PhD degree in Geography and Women's Studies at Penn State University. Her dissertation investigates the role of the drug trade in producing social networks of support, care, power and violence in urban neighbourhoods. More recently, her work extends to considering the carceral geographies of networks of care and the geopolitics of detention in the United States. Her work draws on an ongoing, intimate relationship to Grays Ferry, Philadelphia and feminist research methodologies to understand the complex lived practices of marginal urban spaces in the United States.

Kacy McKinney (BA, MA, PhD) is an Assistant Professor of Interdisciplinary Studies at Marylhurst University, Oregon, USA. She trained as a human geographer and teaches in the areas of international development, South Asia, food geographies, youth, labour, political ecology, graphic novels and qualitative research methodologies. She is also a graphic novelist. Her research focuses on how marginalized children and young people experience shifting landscapes of work and agricultural production in India and Brazil, particularly in the context of the global expansion of agricultural biotechnology. Her work on this topic has been published in the *Journal of Peasant Studies* and *ACME*. A second key area of her research involves the uses of the graphic novel as a tool for a critical pedagogy of geographical theory. She is co-author of *Medusa* (with Milissa Orzolek, She Was Solitary, 2012), a graphic novel about jellyfish. She is currently working on a book manuscript that bridges her artwork and fieldwork in India in the form of a comics ethnography.

Zoë A. Meletis (BA, MScPl, PhD) is an Associate Professor in Geography, and Natural Resources and Environmental Studies, at the University of Northern British Columbia, in Prince George, BC, Canada. Her academic background includes anthropology, environmental studies, planning and 'environment'. She is a human geographer of the environment, who draws upon various critical literatures (e.g. critical conservation studies; environmental justice; political ecology). Her dissertation and related projects consider differences between ecotourism and conservation in theory vs. practice. She has contributed to several projects in Costa Rica, the United States and Canada, on local perceptions of environment, development and change with respect to coastal development, tourism development, environmental management, fisheries, cultural preservation/loss and solid waste management. She is particularly interested in relationships between official claims made and the opinions, views and stories of members of groups 'spoken for' and/or lacking representation. Her research and teaching emphasize explorations of access and ownership, power, discourse, gender and justice. She enjoys collaborative work, and has been published in journals such as *Leisure Studies*, *GeoJournal*, *Journal of Rural Studies*, *Geography Compass* and *Antipode*. She regularly tries to offer her students and readers elements of hope, so that they do not get overly 'eco-depressed'.

Pamela Moss (BA, MA, PhD) is a Professor in Human and Social Development at the University of Victoria, British Columbia, Canada. She trained as a geographer and teaches in the areas of social justice and praxis. Her research spans diverse topics, such as autobiographical writing as analysis, innovative and bodily feminist methodologies, and the material and discursive effects of various subjects including traumatized soldiers and women living with fatigue. She takes up the themes of experience, power and method in all her publications and tends to focus on demonstrating how concepts can explain what is going on in some aspect of daily life.

Maureen Sioh (BSc, MSc, PhD) is an Associate Professor in the Department of Geography at DePaul University, USA. She trained as a hydrologist and has worked in China, Thailand, Cambodia, Malaysia and Singapore and with First Nations communities in Canada. Her research focussed on erosion, deforestation and agriculture before moving on to mathematical modelling of industrial and urban pollution. Concerned with the subjective process of how communities made decisions when faced with environmental issues, her research evolved to postcolonialism, and then to psychoanalysis and economics. She has published on the geographies of anxiety and territorialization and, more recently, on psychoanalysis, sovereignty and foreign investment.

Maral Sotoudehnia (BA, MA) is currently a PhD Student in the Department of Geography at the University of Victoria, Canada. Her research interests include anything that involves culture, people and place. Her dissertation research examines the cultural reception of fictitious capital through various forms of toponymic commodification, including cryptocurrencies. She situates her work at the theoretical intersection of structural and poststructural approaches to economics. She has presented at regional and international conferences.

Ebru Ustundag (BSc, MSc, PhD) is an Associate Professor in the Department of Geography at Brock University, Canada. She is also affiliated with inter-disciplinary MA program Social Justice and Equity Studies. She is a feminist urban geographer interested in exploring various forms of citizenship practices with a specific emphasis on materiality of everyday life. Her research interests include feminist methodologies, community based research, geographies of sex-work, geographies of health and addiction, as well as scholarly activism and research as resistance. Her current research is based on understandings of microgeographies of street-level female sex workers in St Catharines, Ontario, Canada.

Acknowledgements

We appreciate the enthusiasm with which our colleagues and students responded to our call to examine intimacy as part of feminist geography research. Such excitement buoyed us as we wrote, edited, revised and edited some more. We sustained our commitment through sad times around illness and joyful ones around births. We are happy to say that the work has not kept us from maintaining the intimacies in our personal and professional lives.

Katy Crossan at Ashgate Publishing was an ardent supporter of the book from its conception. We were sad to see her leave the project. We wish her well. Amanda Buxton, also at Ashgate, ushered us through the administrative processes admirably. Faye Leerink took over the book project at Routledge and seamlessly transferred the project to the new publishing house. Priscilla Corbett made the production process seem easy, although we know that this is not always the case. Preparing the manuscript took much work from many people. We thank Hannah Moss for preparing the manuscript for our submission. Her keen eye and persistence has no doubt made the manuscript more readable. We thank Tim Rutherford-Johnson for his close read of the text and for picking up all the things that we had trouble deciding about (especially the bibliography) or had simply just missed. We also thank Cameron Duder for developing the index. By reading the text through various terms, concepts and phrases, his index makes the disparate chapters come together in subtle but important ways.

I, Pamela, wish to express my heart-felt gratitude for my friendship with Jason. Our shared intimacies mean so much. From him I continue to remember to fall in love with the everyday. I also am looking forward to watching Larissa and Bennie Charlotte generate pathways through which they can make their ways through the world.

I, Courtney, would like to thank Pamela, who is a joy to work with and an incredible inspiration. You have set an example of excellence as a researcher, mentor and role model. I also want to express my gratitude and love for my husband, Ian Duncan. Every time I said I couldn't, you reminded me that I indeed, could. I am excited for the next adventures we pursue with beautiful Oona Teresa.

Pamela Moss
Willows, Oak Bay, British Columbia
August, 2016

Courtney Donovan
Oakland, California
August, 2016

Introduction

Introducción

1   Muddling intimacy methodologically

Courtney Donovan and Pamela Moss

Intimacy takes many forms. It can be a long-term emotional connection with, a familiar awareness of, or a deep attachment to someone or something. It could relate to a life partner. A colleague. A child. A knowledge. An image. A colour. A piece of music. A sensation. A fragrance. A distant thought. Connections, awareness and attachments have varying intensities, tempos and arrangements. Intimacy emerges through affect, feeling and sensation. It can be an ambient blushing, settling around your shoulders while sitting in a coffee shop on a stormy afternoon, even if only among strangers. It can be a conversation via hastily exchanged words, emoticons, hand gestures or bodily movements, the effects of which languish throughout the day. It can manifest in a shared understanding, communicated in a seeming secret code between people, across species, with technology or amid nature. It can be weekday morning sex before waking the kids to get ready for school. It can be a daily, mid-morning swim in a public pool, lap after lap after lap. Though often cultivated through setting a scene or creating an atmosphere, intimacy can unexpectedly pop up through a chance encounter at a party, on a bus or in a corridor. And, while not singularly emotional, there is an element of sentiment, as light as a brief impression or as strong as a deep-seated bond.

As easy as it may be to identify intimacy, it may be just as difficult to have a conversation about how intimacy can be positioned in feminist geography research. For feminist geographers, singling out instances of intimacies and writing about them is a tricky proposition. First, even though feminist geographers are familiar with data collection methods that are sensitive to the vulnerabilities exposed in the life stories told, documented and analysed in research, that thought of offering one's own intimacies to peers, colleagues and critics can be unnerving personally and professionally. In a highly competitive and somewhat unforgiving environment of academia, exposing a vulnerability (even in support of a wider political project) may weaken a feminist geographer's institutional position and leave her defenceless in the face of assessment audits based on conventional understandings of what counts as research (Mountz, et al., 2015). Second, the personal is political is no longer a unifying principle around which to organize a feminist politics. Feminists, no matter the ilk, are engaged in a political project aimed at transforming society by contesting and disrupting power relations and unravelling their effects. While some feminist geographers claim writing the intensely personal is an indulgent, privileged practice that undermines a systematic dismantling of systemic oppression (e.g. Kobayashi, 2009; Sharp and Dowler, 2011), others find drawing out links between personal experience and wider processes to be useful in contesting multi-scaled arrangements of power (e.g. Moss, 2001a; Butz and Besio, 2004).

These cautions stem from longstanding concerns over what practices feminist geographers need to engage as part of the discipline of geography. A feminist politics manifests not only in the organization and critique of academic workplaces, but also in the intellectual content of feminist geography, especially in terms of methodology. Critical considerations of reflexivity and positionality have spawned discussions over the effectiveness of using personal experience as part of research, resulting in part from the further development of autobiography and autoethnography in geography. The emergence of emotional geographies over the past two decades has paved the way for feminists to extend discussion of affect and emotion into the various aspects of the research process. Thus, it makes sense to locate a discussion of writing intimacy – in its assortment of expressions – at the intersection of these two sub-disciplines.

Acknowledging that intimacy is bound up in the moments that make us feminist geographers – that is, 'who we are, what we do and how we do it' (Moss and Falconer Al-Hindi, 2014) – is the departure point for this collection. Contributors have garnered the verve arising out of living the varied intimacies to show how cracks, patinas and fissures texture how we think about intimacy as a concept, how we approach research in an intimate way and how we do analysis of intimate moments. Writing intimacy is not merely putting a piece of one's personality, identity or subjectivity on display, although such disclosures fill the pages that follow. Rather, writing intimacy is about finessing connections, awareness and attachments to things including each other in order to support a wider political project in a manner that resonates with what we find in our research. Capturing what intimacy is means opening up discussion about what it is, where it exists and how it gets taken up. Grasping the effects of power in particular places means picking apart the research process. Undertaking research responsive to sensate embodiment means figuring out some of the affective, emotional and practical dimensions of intimacy. Writing as an analytical method especially warrants more attention, for it is through writing (which can be an intimate act in itself) that intimacies of everyday living may be brought into dialogue with how to access and work with emotions and affect ethically and politically, in research design, in the field and in the write-up.

By way of introducing writing intimacy into feminist geography, we offer comments on the ways that feminist geographers are conceptualizing intimacy and the implications of these conceptualizations on research. With regard to conceptualizing intimacy, we first lay out what can be termed the intimate turn in geography. We next develop a working definition of intimacy, drawing on the work by feminists across geographies. Our conception of intimacy is grounded in a relational, generative ontology that informs a particular sensibility of research and writing. We then work through how this conceptualization of intimacy might play out within a specific theoretical orientation. With attention to the ways in which power produces subjects, we trace a handful of intimate acts that link some discursive and material aspects of bodies and the places they move through. With regard to the implications of addressing intimacy in feminist geography research, we outline the potential intimacy has in enriching feminist methodology in geography. Our discussion extends writing intimacy beyond personal writing and autobiography into other forms of writing and other places in the research process. We include a discussion of the practices that comprise writing intimacy into feminist geography. Our understanding of the manifestation of intimacy in feminist geography research includes engaging in acts of research, gathering intimate information about research participants, generating data through telling personal stories and exploring the realm of the intimate among researchers and research participants. We close with a brief history of the collection, a rationale for its organization and a description of the book.

The intimate turn

Conceptually, intimacy has emerged across various disciplines as a topic worthy of investigation. Research in fields of study as diverse as business, information studies, rehabilitation sciences, child and youth care, nursing and linguistics have joined the humanities and social sciences in taking up intimacy as a topic of inquiry (e.g. Kivits, 2009; Fritsch, 2010; Pacini-Ketchabaw, Kummen and Thompson, 2010; Jamieson, 2011; East and Hutchinson, 2013; Cahir and Lloyd, 2015). There is literature on the cultivation of family intimacy that might inform understandings of domesticity, such as, for example, women working as nannies through local chains of care, families on vacation seeking historical belonging and parents living in homeless shelters (Schultz-Krohn, 2004; Dimova, et al., 2015; Rojas Gaviria, 2016). In addition, the rise of cell phone use alongside digital communication technologies is creating technology-based social networks that challenge the role proximity plays in intimate conversations and relationships (Hassenzahl, et al., 2012; Jamieson, 2013; Seol, Kim and Baik, 2016). Intimacy is a vibrant concept, with taken-for-granted elements around privacy and humanness being disrupted through ideas about public intimacy (e.g. Stokoe, 2006; Felton, 2013), cultural similarities (e.g. Fortier, 2007; Jankowiak, 2008), and nonhuman and cross-species intimacies (e.g. McHugh, 2011; Wilkie, 2015).

There are constructive ways that feminists are taking up intimacy conceptually, theoretically and methodologically that can inform how feminist geographers might write intimacy. Conceptually, Berlant (2011), an English literature professor, lays out the complex terrain of affective attachments to those things that impede one's own prosperity as a person, something she refers to as cruel optimism. Cruel optimism bears the marks of the intimate, in its range of positive and negative effects. Throughout her work Berlant upends conventional presumptions of what intimacy entails, that of proximity, personal closeness and reciprocity, and replaces them with discussions of the affective expectations of collectivity, sociality and fantasy. She cautions against releasing any optimistic view one seeks of a good life because to do so would mean to forego the necessary change that the world urgently needs. Theoretically, Ahmed (2014), a cultural studies theorist, mobilizes intimacy as part of her theory on the willfulness of subjects. She argues that when there is an error of will, that is when a willful subject fails to attain one's will, there exists unhappiness. Yet if one were to uncouple and re-think the intimate relationship between willfulness and unhappiness, a queering of will could take place. Queering for her is a theoretical practice that, for example, detaches forms of intimate connections and reattaches them to constructs other than that which they are associated, thus producing accounts of the world that change the way one can think and act. Methodologically, Nxumalo (2016), a child and youth care scholar, uses the notion of reconfiguring presence as a strategy to unpick how colonial relations have shaped the understandings of the environment and the subjectivities of those living in settler states. She untwines the intimate connections of the often unexplored colonial pasts with everyday encounters, such as nature walks in childcare settings. She troubles the binary of absence and presence so as not to exchange one for the other; in her aspiration to seek out the bindings of the past to the present, she accentuates 'complexity, uncertainty, contingency and partiality' (p.9).

Feminist geography has had an intermittent engagement with intimacy over the years. Debates about identity, space, place and gender dominated discussions in feminist geography throughout the 1980s and into the 1990s (e.g. Norwood and Monk, 1984; Women and Geography Study Group [WGSG], 1984; Massey, 1994). Interests among feminist geographers tended to poke at the dichotomies of the private and public, production and reproduction, and culture and nature through studies in social reproduction, domestic space, communities and bodies (McDowell, 1999). In addition, feminist geographers debated how research as a feminist differed from mainstream geography, and how such research could contribute to challenging orthodox knowledge (e.g. *Canadian Geographer*, 1993; *Professional Geographer*, 1994, 1995). Many, if not most, of these works tried to link some aspect of the personal or private dimensions of life with overarching structural relations that set parameters around singular identities and experiences. Gendering became an analytical description of a process through which experiences of being woman were integrated into existing analyses of, for example, labour, the state and the city. In a sense, one could understand these works as pursuing the intimate and its links to wider structural processes and relations of power.

Yet the specifics of intimacy have transformed into the sites through which contemporary analysts problematize experience theoretically and methodologically. Slocum (2008), for example, understands intimacy as a dimension of embodiment through which the materialities of bodies come to matter in relationship to other dimensions of embodiment, such as race, in public practices of growing, preparing, selling and eating food. In contrast Gökarıksel (2012) and Smith (2012) use the backdrop of geopolitics to think about intimacy. Gökarıksel (2012) demonstrates how women's bodies, through the intimacies of veiling in public at various places across the city, both define and disrupt the polarizing politics of secularism and the Islamic in Istanbul, Turkey. Smith (2016) investigates the flipside of denoting the effects of intimacies and carves out a pathway – from her research on intimacy as a topic to her own negotiation of intimate relationships while doing research – to integrate insights about intimacy back into academia in order to facilitate a more informed reading of geopolitics.

Feminist geographies have also been part and parcel of the emergence of emotional geographies in the past couple of decades within which studies of intimacy can be situated. Embodiment has been important in examining spatialities of emotion and affect, including scale, proximity and governance (e.g. Thien, 2007; Pain, 2009; Oswin and Olund, 2010). Yet recent research has shown that the body is not the only site at which intimate acts take place, nor is it the only mediator of intimacy (see, e.g., Friedman, 2005; Hayes-Conroy and Hayes-Conroy, 2008; Tamas, 2014). Feminist geographers have brought into focus the ways in which emotional geographies operate in the context of the environments that frame everyday activities and interactions. For example, Jayne, Valentine and Holloway (2010) draw out paradoxes in the idea of intimate social relations in their exploration of the embodied and affective nature of drinking in everyday life. And Lees and Baxter (2011) show how fear is wrapped up and integral to the experience of the arrangements in and elements of a council tower. Investigations of related emotions that take up issues of intimacy try to figure out the spatial implications of particular emotions, such as, for example, the various ways love is connected to place formation and spatialities (Morrison, Johnston and Longhurst, 2013). Methodologically, emotions have been central to elaborating ways to do research, including hate and participatory research, insecurity and autobiography, and joy and collective biography (Cahill, 2010; Bondi, 2014a; Kern, et al., 2014). These kinds of works that relate to intimacy as an emotion or part of an affective assemblage speak directly to the private, personal and innermost attributes of individual lives that feminist geographers have been interested in for some time (after Simonsen, 2007; Colls, 2012).

Intimacy has been addressed most extensively within feminist geopolitics through the concept of the global intimate. Feminists have worked diligently over the past two decades to establish intimacy, expressed as the banal fragments of everyday life, as a politicized entity. As part of a critique of the political and economic processes comprising globalization, feminists have systematically unsettled the artificial boundaries between the global and local, the public and private, and the state and the family (Mountz and Hyndman, 2006). Feminists have also brought into sharp relief the materiality of the relations of social reproduction, which cannot simply be reduced to transnational capitalization (Katz, 2001). Analyses of the global intimate seek to link various actors within global and transnational networks and trace their effects on, for example, place-making, violence, domestic labour and state formation (Dyck, 2005; Wright, 2008; Pratt and Rosner, 2012; Pain, 2014). Within feminist geopolitics, Pain and Staeheli (2014, p.345) define intimacy as three sets of intersecting relations: spatial relations, a mode of interaction and a set of practices. All these relations draw out proximity and the personal as their central features – connecting proximate bodies and households with distant ones within 'interpersonal, institutional and national realms' (p.345).

Yet intimacy as a more inclusive concept, one that includes those things that Pain and Staeheli (2014, p.344) exclude, has yet to be articulated in feminist geography, or in geography more widely for that matter (see Peterson, 2016). Intimacy, including the self, the personal, the proximate, the autobiographical, subjectivity, writing, artistic expression, inanimate things and nonhuman beings as part of the geographies closest in (Longhurst, 1994), could be a concept around which feminist geographers might organize the types of issues, topics and interests feminist take up within the sub-discipline. Valentine (2008) lays out an empirically-based organization of studies related to intimacy within geography. She maintains that the work in sexuality, family studies and children's and youths' geographies is about intimate relations and affective structures. She refers to a 'private turn' (p.2106) within the discipline wherein each of these fields of study address the 'complex inter-spatiality' (p.2097) of the everyday while challenging the false dichotomy of the public and private. If thinking about intimacy in Valentine's terms – as a "specific sort of knowing, loving and caring for a person" that can embrace not just sexual and parenting relationships but also forms of care and affective structures including friendship' (p.2106; after Jamieson, 1989) – then geographies of care and love might also be part of her personal turn (Atkinson, Lawson and Wiles, 2011; Morrison, Johnston and Longhurst, 2013).

In contrast to the production of disciplinary knowledge that Valentine locates her arguments, Peterson (2016) calls for queering a specific concept used within a field of study: the globally intimate in feminist geopolitics. She makes the point that while feminists have excelled in weaving intimacy into geopolitical analyses as a set of sites, spaces and scales, there are still mechanisms involving intimacy that only studies in sexuality, affect and intersectionality can get at. Because intimacy is often linked to sexuality, queer studies is well-placed for an analysis of the globally intimate in geopolitics. She argues that queering the globally intimate by 'deconstructing or "making strange" what appears as "normal" or the "natural order of things" provides critical leverage with no recognized parallel in feminisms' (Peterson, 2016, n.p., first page). This type of queering can both confront the challenges that difference and identity pose (via postcolonialism) as well as track some of the effects of emotion and subjectivity in human behaviour.

Price (2014, pp.509–10), in a discussion of one particular field of study, that of cultural geography, talks about the 'cultural turn' of the 1990s giving way to the 'intimate turn' in the 2000s. Price reads this jarring and somewhat tempestuous process as one where she was trying – much like many feminists at the time, though 'precious few' – to fit 'neighborhoods, bodies, and the inner worlds of human experience and cognition' into political and economic structures (p.507). She argues that this intimate turn is a 'relational, affective, sensual, dynamic, and non-representational approach that views the emergent and the partial as resonant' with the 'consciously embodied self is folded within; emerging inseparably from what was heretofore theory's object (culture, but also place and landscape)' (p.510). In her overview of the sub-discipline, where micro-scale accounts are as commonplace as personal, embodied and emotional ones, she insists on not forgetting the lessons of the past and cautions those engaged in the intimate turn to remember that the purpose of personal storytelling as a way to link everyday things to geography scholarship and that the urgency of political change demands that change still be integral to geography's agenda (pp.515–17).

The intimate turn widely conceived concerns multiple fields of study, scales of inquiry and knowledge production practices. For feminist geographers, this intimate turn may not be as earth-shattering as in other parts of the discipline. Feminist geographers have been attuned to the private, the personal, the body, the emotional, the particular and the intimate long before non-feminist geographers had taken up intimacy as a scholarly issue. Indeed concerns of the everyday shape feminist understandings of how the world works, socially, culturally, economically and politically, with intimacy integral to each act, event and encounter. A contribution to disparate discussions about intimacy – as a site of inquiry, a space of assemblage and a scale of investigation – could bring into focus the methodological issues associated with intimacy in research. Accessing intimacy in research involves a range of conceptual, methodological, analytical and ethical matters. By way of writing intimacy into feminist geography, we now turn to drawing out elements of intimacy as a concept to facilitate a discussion of what feminist geographers are actually looking at in the field.

Conceptualizing intimacy

Given the range of definitions of intimacy feminist geographers are working with, we are somewhat at a loss as to where to begin framing this collection on writing intimacy into feminist geography. An overarching framing of intimacy, something like what Valentine (2008) offers, is useful in claiming intellectual disciplinary space, something we definitely support. Opening up the definition of intimacy highlights features of intimacy that work to explain aspects of a particular topic, as for example Pain and Staeheli (2014) emphasizing spatiality, interaction and practice to capture the process of politicizing economic and political relations through individuals. Yet neither really fits our goals. We have found that a delineation of what counts as intimate, in a concrete and commonsensical way is useful in challenging hierarchies of conventional knowledge by showing how enmeshed the intimate is in multiple realms of life.

As talking points, these conceptualizations can facilitate a discussion of what draws feminist geographers to intimacy. Feminist geographers continue to emphasize the personal as something that exposes the innermost aspects of individual lives. Attentiveness to the banal, mundane practices comprising common everyday activities discloses the unremarkable as a site worthy of investigation. Banalities yield insight into processes that exalt some formations over others, such as formation of a specific collective identity igniting social change (Vacchelli, 2011) or the linking of state security measures with tourism to solidify militarization as a prominent state formation process (Ojeda, 2013). Feminist geographers remain committed to sorting through myriad influences and effects of knowledge production, whether in terms of disciplinary knowledge (spatial, place-based and geographical) or of everyday mobilizations of what counts as being valuable enough to know. Variation in what is available to draw on regarding knowledge – in both form (assumptions, texts, protocols, discourses, etc.) and content (friendship, motherhood, gender, intimacy, etc.) – creates uneven topographies across social difference. Teasing out how place matters and how spatialities coalesce involves tracing the effects of particular knowledges on, for example, the banalities comprising specific sets of activities (on delivery of mental health care on the battlefield, see Moss 2014a, and on resisting neoliberal subjects in climate governance see Bee, Rice and Trauger, 2015). Feminist geographers maintain that there is a political dimension to everyday acts. Indeed, politicization has been a prominent topic in debates over the past several decades, particularly with respect to knowledge production and methodology (e.g. Moss, 1993; WGSG, 1997[2013]; Moss, 2002; Moss and Falconer Al-Hindi, 2008; Laliberté and Schurr, 2016). Although some topics segue easily into the field of politics (e.g. the city, global economies, state security), other topics are less obviously political (e.g. emotions, embodiment, experience), but nonetheless inform a feminist political praxis.

Bearing in mind a commitment to the personal, knowledge production and politics within this intimate turn, we have come to realize the usefulness of laying out some of the premises upon which we have come to base our notion of intimacy. Refusing the triviality of both routinized acts and those acts that appear to arise from nowhere opens up yet another point of entry point into understanding embodied ways of knowing and action, including things like emotion, affect, physical sensations, kinetics, proprioception and changes in comportment. Yet there really is a nowhere when it comes to banality. Parallel to Haraway's (1988) use of the god-trick to signal the existence of an all-knowing subject positioning, inquiry into the personal means dismantling that which sets up the personal to be personal. Acknowledging intuition, gut reaction and affect tends toward a conceptual framing of nonrepresentation (Cadman, 2009), but this need not be the case. The bodily matters and is inextricably wrapped up in both materiality and representation (Braidotti, 2011). Thus taking the personal seriously as part of an inquiry into intimacy in feminist geography is not merely an update to the personal-is-political tenet of feminism; rather, it parallels and is an extension of the geographies that are closest in (Longhurst, 1994).

For us exploring intimacy, intimate acts and intimate writing entails recognizing the interconnectedness of individuals and nonhuman entities, including nonhuman living things and organizations of sets of relations, such as structures of inequality and institutions. In this respect, an 'affective sense of identification' (Milne, 2014, p.214) exists that is fuelled by elements of shared knowledges, whether emergent from a single entity (e.g. critical reflection), between things, or as part of a conventional, perhaps even normalized, discursive formation (Mills, 1997). These knowledges are not fixed, immobile or reliant on given truths; they are productive in that they constitute things – both in the abstract, such as an intimate encounter, and in specific enactments of that abstraction. As part of exploring intimacy and its various forms, we maintain that the production of knowledge includes intimacy as a series of acts that are part of a process of becoming meaningful both individually and collectively. As researchers, we sort through discursive fields (Friedman, 2005) and the materiality of the constitution of subjects (Hekman, 2010) in order to make claims about the links between intimacy, knowledge production and the personal. Thus intimacy, intimate acts and intimate writing are ways that individuals not only reveal the concrete, individualized self, but also exhibit particular characteristics that inform a set of experiences collectively (after Linke, 2011). Indeed the distinctions in how individuals and groups of individuals through identities and subject positionings share and acknowledge the experience of intimacy is significant in mediating the social processes that organize and structure society.

Although attending to the emotional aspects of intimacy is important, reducing intimacy to a singular emotion undermines the salience of intimacy in connections between and among people and nonhuman entities. Relationality means holding in tension the various things that generate intimacy both as an emotion and as a descriptor of a connection or relationship. Unlike understandings of connectedness and relationality as blissful, utopic and Pollyanna-ish, we think, like Berlant (1998), that intimacy need not be considered as secure or even positive. Intimate acts of self-disclosure, including intimacy and intimate writing, are precarious and at times uncomfortable, unpleasant and perhaps even violent. Thus, we seek to write intimacy into feminist geography with the understanding that it relies on close connections defined by a range of characteristics that are fluid, porous, adaptable and in flux – with no inherent predilection towards any one feeling for the specificity of emotion and sensation arises out of the context within which it emerges. This understanding permits a range of affirmative, generative possibilities that can open up the potential of what an inquiry into intimacy can offer.

Our purpose here in suggesting a set of premises from which to think of intimacy is to prevent intimacy from being reduced to or submerged within another concept, such as, for example, emotion, a social relation, a personal relationship or even the everyday. Within the intimate turn, we maintain that the unremarkable alongside the extraordinary can spawn insight into how intimacy works, what it does and what it produces. In other words, intimacy is not just content. Expressions of intimacy are temporally and spatially sensitive. They may be condensed, stretched, pocked with absences or simply scrunched up and pushed aside. And they may even be tinged with the discursive and material histories of those involved. It also may be that their effects linger for but a second or be embedded into a lifelong memory, perhaps recallable at some point in the future or submerged somewhat surreptitiously in the body. Their presence, however, is unmistakable.

But what of analysis? When the inconstant, generative content of intimacy varies so widely and its form manifests so diversely, how does one go about figuring out what intimacy can do? We find it useful to keep at the forefront of any account the notion of a relational, corporeal ontology (Braidotti, 2011). Because we understand intimacy as generative and productive, its form and expression are inevitably contingent, its specificity singular. If we follow through with relationality, positivity and specificity as premises to understand intimacy, then we can for example distinguish how intimacy works in the generation of *the* subject as well as the generation of *a* subject. Intimacy in analysis as well as analysis about intimacy are associated with disclosure, exposure and enclosure of the mutable yet embodied ways in how the self becomes *a* subject. Because we see that intimacy is entwined in knowledge production and has a politics, we can work towards tracing the mobilizations of discursive-material relations through particular enactments of things comprising our worlds. The banality, mundaneness and everydayness of gestures, movements and practices reinforce the idea that both the fostered and unexpected intimacy of particular moments, acts and events continually shape, move and transform the contexts within which people, nonhuman beings and non-living things, including arrangements of sets of relations, are produced as entities.

Muddling intimacy in research

To date methods to access as well as to delve into intimacy within the research process have not been identified in the feminist geography methodological literature. But this does not mean that they do not exist. There is a recognition that accessing emotions empirically is problematic (Davies and Dwyer, 2007; see also Bondi, 2014b) as is maintaining an embodied analysis (Moss, 2005; Simonsen, 2007). Getting at how intimacy transpires in and affects everyday surroundings is central to writing intimacy into feminist geography. We think of this process of access and attention as muddling. Muddling comprises both the entanglement and disentanglement of things, encounters, assemblages, event, acts and processes with each other that exposes various networks of connections and relations. Muddling intimacy in research means attending to sensations in the body, intensities of feeling, resonance between entities and connections among people, nonhuman beings and non-living things. Attempting to account for the vast array of how intimacy is muddled within the banalities of life is a task just shy of nailing jelly to a wall. Its slipperiness makes it a challenge, but the idea that it matters makes the task worthwhile. Communicating how multiple facets of intimacy are layered into the research process can help sort through how intimacy works, what it can do and what it produces.

One particular research practice, personal writing, has been a popular way among feminist geographers to muddle intimacy in research. Personal writing draws on experience to make an argument about a range of things. Indeed this collection recommends a range of ways to use personal writing to take up intimacy in feminist geography through writing that is neither self-absorbed navel-gazing, nor politically irrelevant. Whether an account of one's own experiences of intimacy, an intimate reflection on doing research or a text about an aspect of intimacy in a research project, writings about intimacy are muddled with layers of intricate analyses of widely-ensconced patterns of institutional, societal and cultural norms and relations. This muddling as both a process and a site of inquiry is important: its exploration can provide insight into how systemic processes play out in specific settings, show what sorts of things influence that specificity, and uncover new, innovative or novel sites for political intervention. Muddling understandings of intimacy into feminist methodologies requires detailing how researchers situate intimacy, intimate acts and sites of intimacies into research itself.

Another way to muddle intimacy in research is attending to the relationality of intimate acts for they raise questions about the constitution and understanding of the self vis-à-vis knowledge production. A number of scholars outside feminist geography who have written about intimacy argue that intimate acts need to be framed as acts with epistemological relevance (e.g. Berlant, 1998; Linke, 2011). Despite the varied forms it takes, intimacy plays a central role in constituting particular knowledges that are present in encounters between people and between people and nonhuman beings and things, including for example households, institutions of state power and kinship ties. Sometimes the intimacy emerging through such interaction needs recognition from groups of individuals, communities or the collective in order to make sense of or even validate these intense interpersonal relations (Friedman, 2005). In this respect, intimacy reflects a particular sensibility of and approach to producing and circulating specific knowledges. Without for example the intimate acts involved in child rearing, sexual relations and social reproduction, the productive power of wider knowledges situated in science (e.g. parenting practices), politics (e.g. anti-LGBTQ laws) and culture (e.g. what counts as a suitable marriage) would diminish.

Drawing out the epistemological aspects of intimacy as an act of muddling recognizes the influence of identity in the circulation of knowledge (see Gilmore, 1994). Feminist scholarship has a long history of how gender and other dimensions of social identity influence knowledge in both its form and content. Indeed, feminist geographers have shown the crucial productive role identity has played in shaping perspectives and situating the experiences of individuals and groups of individuals in relation to, for example, embodiment, food, social reproduction and natural and ecological disasters (see Katz, 2008; Kobayashi and Peake, 2008; Hayes-Conroy and Martin, 2011; Dominey-Howes, Gorman-Murray and McKinnon, 2014; Barber, 2015). Identity produces a specific sort of knowledge that arises out of interpersonal entwinements and everyday experiences and the insight gleaned from the formation of these specific knowledges is contingent on positionality (Haraway, 1988). Through various types of personal writing, the productive (and generative) power of specific knowledges can be useful (Moss, 2001a, 2014b).

Muddling intimacy requires attention to both the unique perspectives of and the disparate linkages among idiosyncratic acts, encounters and events that accompany affective relationships. For example, there is no denying that intimacy brings with it an element of confession, admission and revelation. Foucault (1980b, 2003, 2008) in his work on truth regimes, power and knowledge understood confession to be a central ritual in Western society set up to elicit truth. Though the genealogy of confession can be traced to Christianity, the confession has gained currency in other social arenas, public and private domains alike, including writing (Taylor, 2009). Foucault maintains that the act of confession plays an important role in the process of subjectification, or the construction of oneself as a subject. In other words, he envisions the self as an effect of the negotiation of truth claims revealed through admissions of one's actions by talking about them. The intimacy on display through an act of confession is relational: while confession may motivate an inward evaluation of the self, it is not solitary; confession only matters when there is someone there to listen, to read or to observe (Miller, 2000; Cooke, 2013). Articulations in confessions are usually verbal, but not exclusively. The body speaks through tacit declarations of information that are bound to the knowledges informing observers, who then draw their own conclusions (see Moss and Prince, 2014, 102, on psychiatrists seeking out malingerers). Indeed, the confessional as a mode of communication is relatively open and relies on the unfolding of the interplay between the interlocutor and listener or observer. In this respect, intimate discourses are ones in which an individual or subject is compelled to make known a particular truth claim about the self or her own self. While not revelatory of *the* truth, confessions are actively involved in negotiating *a* truth. Confession is ethical in nature precisely because it requires the presence of another entity (Miller, 1990). This presence for Foucault (1980a) activates power, and when confession is part of a process of jurisprudence or care of the soul, for example, the interaction leads to moral judgement and its associated effects. Yet confession is a technique of the self much wider than acts within structural arrangements of power in Christianity, the state or nature. Confession sets up individuals to regulate their own moral and ethical behaviour, sometimes in public ways (Foucault, 2005, 2011).

A number of studies have identified linkages between self-regulation and personal narrative as part of a set of intimate acts and practices (e.g. Bleakley, 2000; Daniell, 2003; Longhurst, 2012; Fannin, 2013). But is the act of constructing, drawing on or writing the personal necessarily confessional? Lejeune (2009) challenges the notion that intimate writings, including diaries, are by definition revelatory of the self by way of confession. Rak (2009) argues that intimate writing is exclusively the product of power relations and suggest that diaries offer a space for transgressive opportunities and potential for change. Congruent with Foucault's ideas about truth, power and knowledge, they use notions of counter-discourse and counter-memory to distinguish alternative subjects in intimate writings that are not within the purview of surveilling institutions, including society. Foucault (1980c, 2010) accomplishes a similar task in his work on personal, private and intimate documents that speak to the experiences of everyday life. He meticulously demonstrates that these documents and the stories that they convey – both then and now – work to counter prevailing political discourses about crime, sex and citizenship. Through exploiting the vulnerabilities of prevailing discourses, Foucault exculpates confession from the sensational texts for a scandal-seeking audience and sets them up as part of a wider political project contesting prevailing truth regimes.

The muddling of various forms of personal writing to access intimacy in feminist geography research is part of a wider political feminist project. Personal writing is effective at identifying, describing and contesting existing truth regimes that marginalize, oppress and exploit people, nonhuman beings and non-living things. It also opens up pathways to *think* the world differently, which of course heralds the potential to *act* differently. This potential is the effect of an 'affirmative bond that locates the subject [*a* subject] in the flow of relations with multiple others' (Braidotti, 2013, p.50). Being able to engage in a type of personal writing that generates potential through a connectedness with multiple others permits feminist geographers to grasp how intimacy works and figure out what intimacy does.

Methodological enrichment

Feminist methodologies is one of the longest-standing interests in feminist geography (see e.g. Zelinsky, Monk and Hanson 1982; WGSG 1984). Over the past few decades, methodology has come to capture the conceptual imaginations of feminist geographers, especially graduate students learning to conduct their own research projects and junior scholars who have challenged conventional understandings of research – both feminist and non-feminist ones! Although outside the discipline, particularly in sociology, there has been a spate of feminist methods handbooks since 2000 (e.g. Ramazanoğlu with Holland, 2002; Sprague, 2005; Hesse-Biber and Leavy, 2006, 2007) there are few expressly feminist geography ones over the same period of time (Bondi, et al., 2002; Moss, 2002; Moss and Falconer Al-Hindi, 2008). Many of the feminist methodological issues appear in other collections or in stand-alone peer-reviewed articles. There has, however, been a recent surge in interest in re-engaging more systematically with feminist methodologies and bringing out those discussions in print (e.g. Bilio and Hiemstra, 2013; Kohl and McCutcheon, 2015; Laliberté and Schurr, 2016).

What of intimacy in feminist methodologies? Are feminist methodologies in geography inherently intimate? Is muddling intimacy a useful methodological strategy? Or do such approaches to research merely take on intimacy as a topic or site of inquiry? Rather than debate the purpose of feminist methodology in geography, we ask what work can be done around framing intimacy that can inform debates about research. Writing intimacy into feminist geography via muddling requires a shift in understanding how intimacy is part of and actually constitutes research itself. A conceptualization of intimacy that is relational, generative and power-laden may facilitate a more nuanced discussion of key issues in feminist methodology in geography as well as enhance the presentation of the complexity of what feminist geographers are finding in their research. Addressing intimacy methodologically means thinking through the implications of researching intimacy and intimate acts through, for example, reflection on how intimacy enters into research accounts, how intimate moments with the research process inform analysis and how intimate relationships with one's self, each other and things in each other's lives shape how feminist geographers do research. Accessing intimacy through research as both a topic and as an affective process means drawing out some of the innermost and possibly the most intensely well-guarded aspects of someone's life. Sensitivity and thoughtfulness in research design, analysis and circulation of insights are central tenets upon which inquiries into intimacy need to be based, especially given the variable emotional dimensions of banal connections within the everyday and the intensities with which people experience them. Attending to these methodological principles both abstractly and concretely means mobilizing diverse feminist theories through specific concepts in ways that will inform accounts of intimacy and intimate acts as well as intimate accounts of research. Presenting research about intimacy means considerately exploring which methods are most effective in conveying insights gained through the research while diminishing intimacy's significance as a topic or an affective process.

So where do feminist geography methodologies stand in relation to inquiries into intimacy? Methodologically, the idea of examining that which seems to be hidden is salient to studies of intimacies. Hiemstra (in press) proposes the use of periscoping to explain how feminists come to access those things that are usually hidden from view of the researcher. Periscoping is an embodied ethnographic approach that examines the everyday when conventional methods may not be an option. It may include scanning the research landscape from afar, locating something interesting to scrutinize, dropping in for observation and then withdrawing for critical analysis and reflection. Research topics that lend itself to periscoping might be systems of practices in bureaucracies, corporations or institutions that systematically exclude and marginalize women of colour from services, promotions or employment. What is useful in Hiemstra's periscope for studies about intimacy is the idea of looking at seemingly inaccessible things, like processes and structures, through everyday encounters in various places, like offices, waiting rooms and meeting rooms.

Psychoanalytical approaches in feminist methodologies also seek to access that which is hidden from the purview of the researcher – emotions. Bondi (2014b) takes up the issue of trust and rapport in interviewing. She examines the interaction in an interview with Katherine, wherein she explores Katherine's explicitly stated intension, oft repeated, about not wanting to do harm to others as a counsellor. Bondi picks up this response and links it to their relationship as interviewer and interviewee, where as an interviewer Bondi asks whether Katherine was concerned about her trustworthiness as a researcher. Aside from clarification comments, Bondi does not follow up on the emotions associated with Katherine's statement; rather she continues the interview within which Katherine opens up and tells a deeply personal story of her life. Bondi uses the psychoanalytic notion of 'receptive unconscious' (p.48) as a methodological meditation on how rapport may be established during an interview. The receptive unconscious is a way psychoanalysts heighten their sensitivity and awareness to analysands without verbally communicating. Training and practice permit unconscious communication between the two. Bondi suggests that this idea of receptive unconscious can assist in sorting through taken for granted processes taking place within relationships between researchers and participants.

These types of hidden aspects in research are but two ways to access that which remains mysteriously clandestine, purposefully concealed or humbly unfamiliar. Perhaps thinking in terms of what to do with the disclosures of that which was once hidden is the crux of writing intimacy into feminist geography. Thinking in terms of disclosing the everyday, through moments, acts, feelings, thoughts or confessions, can produce sensitive, mindful, perceptive and insightful accounts of intimacy that refuse to be scandalous exposés of personal lives.

Feminist geographers have the skills to produce such accounts. Autobiographical work is more than about examination of the self as a researcher, a researcher's subjectivity and a practice of self-disclosure. Autobiographical writers have adroitly drawn out the relationship between the singularity of individual experience and the wider processes through which individuals are embedded, either through widening understandings of concepts (e.g. Bondi, 2014a, recounts her own angst and anxiety as a way to reconceptualize ontological security and insecurity) or using particular concepts to frame a mélange of experiences (e.g. Besio, 2005, uses representation in stories of sewing machines to make sense of entanglements of colonial power among women in northern Pakistan). Yet being able to organize these disparate works through the theme of intimacy, rather than in contrast through empirical literatures, a sub-discipline or a theoretical project, could possibly deepen understandings feminists have of experience, knowledge and power, and in their relationships to each other methodologically. For example, there are certain elements that distinguish intimate writing from other forms of written work. Intimate writing and more generally writings about intimacy rely on affective voices that aim to engage the reader in particular ways. Indeed, the author of the intimate writing seeks to create a private sensibility through their narrative and close connections with the reader. While the act of producing and consuming intimate writing tends to be most associated with the private sphere, there is still a publicness of intimate writing (Linke, 2011); the writings are on display. While readers drop in and out of the open exhibition, they remain hidden and a little bit distant from the other readers as well as the author as the one drawing them into the writing. It is for this reason that intimate writing, like other acts of intimacy, is defined by an inherent worldliness (Pratt and Rosner, 2012). Keeping in mind the exposed nature of intimate writing, it is important to consider what happens to intimate acts when they are made public and visible.

Writing intimacy in geography means disclosing the everyday through thoughtful accounts of the lives of people, nonhuman beings and non-living things. In writing these accounts, researchers think through the practices of research inquiry that can capture intimacy and intimate acts and then grapple with how best to understand and represent these knowledges of the banalities of life (Gilmore, 1994). Researchers rely on a range of research methods as ways to access and analyse the everyday. Materials get collected through journals, from archives and in interviews as parts of ethnographies, autobiographies and participatory research. Once amassed, materials get analysed through critical reflection, writing, collaboration, drawing and witnessing. From this non-exhaustive list of research methods emerge accounts of intimacy and intimate acts in as many different forms. This collection is an attempt to bring together some of the work about intimacy into methodological discussions within feminist geography.

The book

When the net is cast widely, it becomes obvious that feminist geography has been and continues to be about intimacy, whether as a topic, a scale, an inquiry, a place or an approach to research. Organizing discussions of intimacy around its conceptualization facilitates engagement with an array of practices that make up feminist geographic research. Muddling as part of a feminist research practice includes ethics and political quandaries feminists face while being positioned simultaneously as a teacher, a researcher, a professional and an activist; data collection methods, such as field notes, ethnographic texts, interviews and personal journals; analytical methods, such as reflexivity, critical reflection and positionality as well as theoretical interventions, conceptual framings and data generation; and writing in the sense of the logic of presentation, publication venues and awareness of form, content and process as part of the research process.

The history of this collection has its roots in a set of three sessions we organized for the Annual Meeting of the Association of American Geographers held in Tampa Bay, Florida, in April 2014. Many of those presentations are included in this collection. We extended the invitation to other researchers who had expressed interest in the sessions – either in response to the call for participation early on in the process, at the conference or during the development of the book proposal. We have tried to bring a range of contributors into the project: by stage in career (from students who have just completed a Master's degree through to professors of over 25 years); by theoretical orientation (as feminists from humanist, poststructural, structural and psychoanalytic traditions); and by topic (scaled from the body to the global including, for example, the archive, flows of economic capital, illness and fieldwork).

There are two types of chapters. The shorter ones are focused interventions into a particular topic. They address issues such as what to do with the things that either comprise or shape research but are not included in formal write-ups (Adams-Hutcheson and Longhurst; Alexander; Sotoudehnia) or trouble, extend or develop concepts that deal with various aspects of intimacy (Donovan; Hanrahan; Henkin; Ustundag). The longer ones tend to take up intimacy topically, conceptually or analytically to demonstrate how it works. Some pick apart the practices that comprise the research process (Askins; Besio; Massaro and Cuomo) and some engage with nonhuman beings and non-living things (Gillespie; Meletis and Hawkins). While some reconsider what research methods do in relation to intimate encounters (Falconer Al-Hindi, Moss, Kern and Hawkins; Moss; Sioh), others use intimacy to rethink how to present their research (de Leeuw; Madge; McKinney). Of course, these categorizations are neither exhaustive nor mutually exclusive. We intend them to be heuristic in thinking about how to read them as texts and how they might contribute to ongoing debates in feminist geography about research intimacy.

We organize the book in four sections that bring into focus four different ways intimacy has been taken up by the authors. In the first section authors write about some of the methodological challenges feminist geographers face when dealing with intimacy in research, including issues about reflexivity, positionality, belonging and that which goes unsaid. In the second section authors address the issues that arise from the decision to include one's own story in research, whether as an inquiry into the form of the presentation or as part of the context within which one works as an academic. In the third section the chapters show a range of topical aspects of intimacy that researchers explore, as a practice-oriented objective, as a research practice, as an affective research process and as a nonhuman emotion. In the fourth section authors take up intimacy analytically through five different research methods: witnessing, recording field notes, reading archival material, writing autobiographically and authoring creative non-fiction.

Methodological challenges

Methodologies sensitive to intimacy are informed by theoretical notions that seek to counter the negative effects of oppression and privilege in the work academics do. Situating intimacy and intimate acts in the context of feminism and epistemology attends to the issue of how best to understand, represent and draw from these knowledges (Gilmore, 1994; Ackerly and True, 2008). Reflexivity and self-positioning in the research process are epistemological practices that continue to be challenging. Maral Sotoudehnia explores some personal moments where the contradictions in her identity in the field are exposed. These particular instances – as reflexive, autobiographical accounts drawn out of her field notes – show fluctuations in her hybrid identity that she continually mediated outside the boundaries of her formal research that included interviews with planners and local officials in Dubai. She expresses anxiety and apprehension over her skills to address the slipperiness of her attempt to position herself and to negotiate her own placement by those around her.

There is so much that is part of research that gets written out, ignored or stacked in a pile of notes shoved to the side of a book shelf. Reflecting on what is left out and goes unsaid, Gail Adams-Hutcheson and Robyn Longhurst pull apart their own positionings as a researcher and a supervisor in relation to an unsettling phone call from a potential research participant. They write about how that which is undoubtedly bodied, such as the deep emotional distress over being uprooted from home through an earthquake, remains unsaid in the very research designed to be about displacement. The untidiness of how knowledge emerges does not correspond with the process through which research is planned.

Even though much of the production of feminist geographic knowledge about methodology arises from a collective interest in how researchers as selves are inserted into and as part of the research process, direct accounts of a researcher's entanglement within various sets of intimate relationships are absent. Maureen Sioh, through the metaphor of a fugue, follows some of the threads spun by colonialism, capitalism, gender and race that connect her story with those of elite global financiers. She goes beyond a mere tracing of relationships into the realm of psychocultural mechanisms as part of a system of repetition and connection in financial decision-making. She uses intimacy as a methodological practice to access the embodied affect that engagement between researchers and research participants rely on to understand the topic and one another.

Yet feminist researchers not only scrutinize the research method they use, they also query how to convey these connections to different audiences. Recognizing audiences, particularly academic and community, has long been of interest to feminist researchers (Letherby, 2003). Vanessa A. Massaro and Dana Cuomo take this notion further, and unravel their own complex positionings implicated in the worlds of their research participants and the communities they have affinities with. They delve into some of the intense emotional aspects of their intimate connections with their research participants as they present their research to a range of people in academia and the community. Fielding reactions from academic colleagues of shocked disbelief over access to socially closed groups and of smug skepticism over critical empathy for incarcerating domestic violence offenders, they find a way to bridge audiences through discourse and reflection. Using presentation as methodological encounter, they propose to nurture ethical, inclusive research relationships by facilitating engagement and building empathy across audiences – academics, research participants, community members and professionals.

Emergent effects of including one's own story

Once the decision is made to include one's own story in an analysis of intimacy, two key issues surface around form and content, that is, how to write it up and what to include in one's own story. Various techniques in writing can be used to represent the experience of intimacy, to draw the reader into the text and to create meaningful encounters between both the author and the text, and the text and the reader (see Wyatt, et al., 2010; Kaufmann, 2011; Moss, 2013; Pansoneau-Conway, et al., 2014). Issues of disclosure matter methodologically and analytically: revelation entangles the writer and reader (just as it does in the field between researchers and entities within research), and confession contributes to an authorial voice (even though it makes an analyst vulnerable). The authors of the chapters in this section keep these two issues at the fore of their writing.

Clare Madge, in light of using creative means through which to express geographic analysis, makes an argument that disclosure of one's own intimacies needs different, more innovative forms of presentation and analysis. She offers a multidimensional form of expression to capture her multilayered experiences of two moments in her life: pregnancy and illness. Her photographic and poetic work indicates that there is something else about intimacy that appears in these analytical expressions that cannot be articulated in conventional academic ways. Likewise, Kacy McKinney uses graphic panels to convey intimate field moments analytically. Affect left outside the written word can not only be incorporated but also embraced graphically, giving more oomph to an analysis of the process of disclosure. She accomplishes a double move of muddling: she pulls in the reader through her graphics while telling a confessional story that enmeshes herself and her research assistants into each other's lives.

This muddling, the existence of which is recognized widely, is difficult to disentangle. Tracing connections and untying knots is repetitive work that uncovers links seemingly useful only to a single life. But this is clearly not the case. The contexts within which researchers do research frame (to both foster and limit) how one goes about using this type of autobiographical information within research.

Zoë Meletis and Blake Hawkins acknowledge their place-based identities as part of an encumbrance they carry with them as researchers. They recognize the relationality of their connections to the places where they have lived and the places where they conduct their research. Northern British Columbia is integrated into their everyday lives that affects the way they live, what they choose to research and how they do their academic work. Their chapter about the constitution of research identities is as much an opening for them to begin the disentangling process as an exercise in understanding themselves as an invitation to readers to reflect on their own stories. Toni Alexander takes on a similar task of tracing her own enmeshment in a more specific and less supportive place. She examines the effects of disclosing disabling illness in the workplace and the manipulation of that information in a hostile work environment. Part of disencumbering herself from threatening and harassing behaviour meant reorienting her research interests topically and methodologically; namely, she uses her experience to scrutinize the ways in which individuals engage with governing institutions that shape her work as an academic as part of building a career.

Difficulties lie in trying to figure out how intimacies form not only in the context, places and environments that researchers do academic work, but also in the very way that researchers *do* research. Kye Askins explores how she uses a particular tool, that of doing-writing, to cultivate a set of practices that can assist her in writing about emotion, affect and intimacies as an academic. She draws out the relationality of bodies and emotions by engaging in a set of energy-based exercises (based on *Qigong*) that bring parts of her body into relation with memories, thoughts, feelings and sensations. Her critical engagement with her body becomes her story that she includes in the process of writing about her researcher's self.

Multiple aspects of researching intimacy

Intimacy as a topic of research emerges in different ways. Sometimes the research tries to get at the positive aspects of intimacy that link the personal to feelings of joy, happiness and belonging. Yet there are negative aspects of intimacy that need attention. Courtney Donovan investigates the way in which the claim of accelerated intimacy among digital health communication enthusiasts makes a bid to humanize doctor–patient encounters. She argues that promoting digital medicine through, for example, online surveys, patient portals and teleappointments (via teleconferencing platforms) as the way physicians, hospitals and clinics can involve patients in their own healthcare produces contradictory effects on people seeking healthcare. While convenience matters for busy people, access matters for marginalized people and availability of evidence matters for informed people, neoliberal discourses of individualization and responsibilization of healthcare for individuals go unchallenged.

Attending to the positive and negative effects of emotions in research has opened up new understandings of the conduct of research of both affect and feelings. But what happens when the researcher is constituent of the intimacies within the research itself? Kelsey Hanrahan tells the story of her entwinement into the lives of the people she lived with during her fieldwork. She reflects on her involvement in the caring for a dying woman through the notion of handling, arising from *joo* as the Konkomba word to express the activities of caring for someone. She relates a series of intimate acts that makes the concept of handling as part of caring more fulsome, connected and relational. Through her reflections, she offers the idea of handling as an intimate research practice, the sentiment of which flows beyond care for the dying towards caring more generally. The idea that intimacy gets generated through research is exactly what Karen Falconer Al-Hindi, Pamela Moss, Leslie Kern and Roberta Hawkins explore. Through the methodology of collective biography, they provide an account of a set of reflections about moving cabins when several mice make their presence known. They muddle themselves through their recollections of intimate acts – that significantly differ – leading to their enmeshment as individual researchers and as a research group. They use the term 'becoming collective' to highlight both the intimacies that constitute themselves as researchers and the process intimacy plays in fueling that constitution.

The link between intimacy as an emotion and the intimate acts that lead up to particular research moments (including writing) provides insight into questions about the researcher's self and how researchers relate to other people, non-human beings and non-living things that are part of the research process. Kathryn Gillespie unravels her own embeddedness in the lives of the species that are part of her academic and home life – cows, calves and cats. She uses intimacy as both a topic to investigate and as a research method. She displaces the researcher as the centre of the research process, and invokes an emotional sensibility based on a cultivation of attachment that can carry her forward. Empathy emerges as a research strategy that facilitates access to the emotional worlds of nonhuman life.

Analytical methods as part of writing

While all the authors in the chapters in this fourth section use autobiographical writing to tell their tale, each uses a different analytical method and strategy of writing to weave together the intimacies they encounter in the research process. Through their stories, they open up conceptual and theoretical pathways to deal with their research data, as in a process of collection and as information to analyse. Each author too is caught up in and distant from the intimacies they encounter, whether as emotional bonds, emergent affects or thoughtful cultivations. Samuel Henkin employs witnessing as a method through which to access the lived connection to gay men, that is to say *a* gay man, who survived the Holocaust. He brings the past into the present, *his* present, as a way to remember the effects of extreme violence. By bearing witness he draws out the intimacies that can no longer remain unspeakable – he facilitates the speaking of the unbearable suffering so that he, and society, does not forget. Such politicized notions of intimacy, despite the varied ways in which intimacy manifests, attend to the constitutive configurations of power and knowledge that emerge from mutual recognition and intense interpersonal relations, even if dislodged in time.

Coming to terms with intimacy politically can reflect a type of sensibility, a type of approach to research, that can inform the production of and response to knowledges circulating more widely. Ebru Ustundag traces one dimension of the production of intimate knowledge in her field notes. She notices that she writes in both Turkish, her first language, and English, her academic one. Yet the expected patterns of the personal for reflection and the analytic for research matters did not hold true. In telling her story of her work with street-level sex workers, she begins to question what types of subjects are being produced through her practice of recording her research in a set of written notes. Her work highlights a notion of becoming that gets beyond that of perpetual motion and towards an understanding of the mundane but generative links between encounters, events and records in the production of knowledge.

Sorting through the effects of specific configurations of power can assist a feminist researcher interested in intimacy in understanding the limits of the inter-pretation of data, particularly around subjectivities and subject positionings. Kathryn Besio, in her study of intimate gardens, looks to the archive for informa-tion. She reflexively details her process of her autoethnographic engagement with gardens through a family photo, *Scene in Hassinger Garden*, from the 1890s. Lurking, as the concept through which she reads the photo, is twofold: it describes her methodological practice as a researcher walking the streets of Hilo hoping to talk to people about their gardens as well as her analytical process of being sensi-tive to the gendered, capitalist, colonial relations that marginalize certain subjects. This combination pulls out the complexities of the lived intimacies of identity and subjectivity that have been obscured in the history of Hawai'i.

Even though authors expose their own and others' private lives in writing up their research, there is still ongoing mediation of what can and cannot be revealed through the writing. Autobiographical writing as analysis uses the self as a depar-ture point from which to tell stories that are about more than just a self. Pamela Moss traces some of the micro-connections within biomedical treatment decisions in an intensive care unit. Her story, organized around her experiences of trying to make decisions about her mother as her mother's legal medical voice, highlights what happens when input about care via family members of dying relatives has been closed off because of the heavy reliance on texts in the practice of medicine. Her story is also a decidedly personal one, one that features tedious minutiae of events and actions. Yet when taken together, these specifics reveal the limited options, via subject positionings, open to her while engaging with an institution.

Much of what is forgotten about methods of analytical writing that centre on the self is that such writing is less about the self and more about privileging a specific narrative that specifies connections between individual entities and wider structures and processes. Yet this is not how one experiences the banalities of the everyday. Sarah de Leeuw uses creative non-fiction essay writing as a means through which to convey insight from her analysis of injustice, colonization and oppression in a form intended to resonate with the everyday of readers. She tells a story that is much like the kind of lived stories that feminist geographers are trying to analyse. By sculpt-ing the narrative, she shows how the analyses of everyday life are strung together in ways we do not necessarily see in the moment. She pulls together the bits and pieces that make up the particulars of the quotidian in a text written to do what analyses do: provide insight into what is happening so that there can be change.

Across the pages: An invitation to engage in muddling

Both as a compilation and as individual pieces of work, these chapters attempt to disentangle some of the muddling that makes up research about intimacy among people, nonhuman beings and non-living things. This muddling includes engagement with intimacy in a variety of ways, as for example, a topic, an emotion, a form of self-disclosure and a series of affective moments. The authors also take up intimacy as a constituent factor in analytical processes, in interactions with research partners, participants and texts, and in reflections on intimate acts. Across the pages there are examples of how to write intimacy into feminist geography. By laying bare parts of the muddle, the authors unravel tightly-bound connections, coiled with distinctive awareness and recognition of unforeseen attachments in, of and through research encounters. No matter the topic, all the chapters are forms of intimate writing and each foregrounds the resonance of a relational, generative and politicized conceptualization of intimacy. They include depictions of how relationality illustrates the ebb and flow of the everyday, how affect shapes the unfolding of a research project, how feelings initiate change in analysis, how sensations necessitate some form of action and how writing exposes knowledge production processes. In short, what these chapters offer is a set of ways to get at intimacy and to figure out what intimacy does and what it can do. And, as a reader, you are most welcome to do the same: muddle intimacy methodologically and entwine your selves into the texts to assist in figuring out what intimacy does and what it can do.

Works cited

Ackerly, Brooke and True, Jacqui, 2008. Reflexivity in practice: power and ethics in feminist research on international relations. *International Studies Review*, 10(4), pp.693–707. doi:10.1111/j.1468-2486.2008.00826.x

Ahmed, Sara, 2014. *Willful subjects*. Durham: Duke University Press.

Atkinson, Sarah, Lawson, Victoria and Wiles, Janine, 2011. Care of the body: spaces of practice. *Social and Cultural Geography*, 12(6), pp.563–72. doi:10.1080/14649365.2011.601238

Barber, Tamsin, 2015. Performing 'Oriental' masculinities: embodied identities among Vietnamese men in London. *Gender, Place and Culture*, 22(3), pp.440–55. doi:10.1080/0966369X.2013.879101

Bee, Beth A., Rice, Jennifer and Trauger, Amy, 2015. A feminist approach to climate change governance: everyday and intimate politics. *Geography Compass*, 9, pp.339–50. doi:10.1111/gec3.12218

Berlant, Lauren, 1998. Intimacy: a special issue introduction. *Critical Inquiry*, 24(2), pp.281–8.

Berlant, Lauren, 2011. *Cruel optimism*. Durham, NC: Duke University Press.

Besio, Kathryn, 2005. Telling stories to hear autoethnography: researching women's lives in northern Pakistan. *Gender, Place and Culture*, 12(3), pp.317–31. doi:10.1080/09663690500202566

Billo, Emily and Hiemstra, Nancy, 2013. Mediating messiness: expanding ideas of flexibility, reflexivity, and embodiment in fieldwork. *Gender, Place and Culture*, 20(3), pp.313–28. doi:10.1080/0966369X.2012.674929

Bleakley, Alan, 2000. Writing with invisible ink: narrative, confessionalism and reflective practice. *Reflective Practice*, 1(1), pp.11–24. doi:10.1080/713693130

Bondi, Liz, Avis, Hannah, Bankey, Ruth, Bingley, Amanda, Davidson, Joyce, Duffy, Rosaleen, Einagel, Victoria Ingrid, Green, Anja-Maaike, Johnston, Lynda, Lilley, Susan, Listerborn, Carina, McEwan, Shonagh, Marshy, Mona, O-Connor, Niamh, Rose, Gillian, Vivat, Bella and Wood, Nichola, 2002. *Subjectivities, knowledges, and feminist geographies: the subjects and ethics of social research.* Lanham, MD: Rowman and Littlefield.

Bondi, Liz, 2014a. Feeling insecure: a personal account in a psychoanalytic voice. *Social and Cultural Geography*, 15(3), pp.332–50. doi:10.1080/14649365.2013.864783

Bondi, Liz, 2014b. Understanding feelings: engaging with unconscious communication and embodied knowledge. *Emotion, Space and Society*, 10(1), pp.44–54. doi:10.1016/j.emospa.2013.03.009

Braidotti, Rosi, 2011. *Nomadic theory: the portable Rosi Braidotti.* New York: Columbia University Press.

Braidotti, Rosi, 2013. *The posthuman.* Cambridge, UK: Polity.

Butz, David and Besio, Kathryn, 2004. The value of ethnography for field research in transcultural settings. *The Professional Geographer,* 56(3), pp.350–60. doi:10.1111/j.0033-0124.2004.05603004.x

Cadman, Louisa, 2009. Nonrepresentational theory/nonrepresentational geographies. In: Rob Kitchin and Nigel Thrift, eds., *International Encyclopedia of Human Geography.* London: Elsevier. pp.456–63.

Cahill, Caitlin, 2010. 'Why do *they* hate *us*?' Reframing immigration through participatory action research. *Area*, 42(2), pp.152–61. doi:10.1111/j.1475-4762.2009.00929.x

Cahir, Jayde and Lloyd, Justine, 2015. 'People just don't care': practices of text messaging in the presence of others. *Media Culture & Society,* 37(5), pp.703–19. doi:10.1177/0163443715577242

Canadian Geographer, 1993. Feminism as method. 37(1), pp.48–61.

Colls, Rachel, 2012. Bodies Touching Bodies: Jenny Saville's over-life-sized paintings and the 'morpho-logics' of fat, female bodies. *Gender Place and Culture*, 19(2), pp.175–92. doi:10.1080/0966369X.2011.573143

Cooke, Jennifer, ed., 2013. *Scenes of intimacy: reading, writing and theorizing contemporary literature.* London: Bloomsbury.

Daniell, Beth, 2003. *A communion of friendship: literacy, spiritual practice, and women in recovery.* Carbondale, IL: Southern Illinois University Press.

Davies, Gail and Dwyer, Claire, 2007. Qualitative methods: are you enchanted or are you alienated? *Progress in Human Geography,* 31(2), pp.257–66. doi:10.1177/0309132507076417

Dimova, Margarita, Hough, Carrie, Kyaa, Kerry and Manji, Ambreena, 2015. Intimacy and inequality: local care chains and paid childcare in Kenya. *Feminist Legal Studies*, 23, pp.167–79. doi:10.1007/s10691-015-9284-6

Dominey-Howes, Dale, Gorman-Murray, Andrew and McKinnon, Sara, 2014. Queering disasters: on the need to account for LGBTI experiences in natural disaster contexts. *Gender, Place and Culture*, 21(7), pp.905–18. doi:10.1080/0966369X.2013.802673

Dyck, Isabel, 2005. Feminist geography, the 'everyday', and local–global relations: hidden spaces of place-making. *Canadian Geographer*, 49(3), pp.233–43. doi:10.1111/j.0008-3658.2005.00092.x

East, Leah and Hutchinson, Marie, 2013. Moving beyond the therapeutic relationship: a selective review of intimacy in the sexual health encounter in nursing practice. *Journal of Clinical Nursing*, 22, pp.3568–76. doi:10.1111/jocn.12247

Fannin, Maria, 2013. The burden of choosing wisely: biopolitics at the beginning of life. *Gender, Place and Culture*, 20(3), pp.273–89. doi:10.1080/0966369X.2012.694355

Felton, Emma, 2013. A/effective connections: mobility, technology and well-being. *Emotion, Space and Society*, 13, pp.9–15. doi:10.1016/j.emospa.2014.09.001

Fortier, Anne-Marie, 2007. Too close for comfort: loving thy neighbor and the management of multicultural intimacies. *Environment and Planning D: Society and Space*, 25, pp.104–19. doi:10.1068/d2404

Foucault, Michel, 1980a. Introduction. In: *Hercule Barbin: being the recently discovered memoirs of a nineteenth century French hermaphrodite*. Translated from French by Richard MacDougall. New York: Pantheon. pp.vii–xvii.

Foucault, Michel, 1980b. The confession of the flesh. In: *Power/knowledge: selected interviews and other writings, 1972–1977*. Edited by Colin Gordon. New York: Pantheon Books. pp.194–228.

Foucault, Michel, 1980c. Truth and power. In: *Power/knowledge: selected interviews and other writings, 1972–1977*. Edited by Colin Gordon. New York: Pantheon Books. pp.109–33.

Foucault, Michel, 1997. *The politics of truth*. New York: Semiotext(e)

Foucault, Michel, 2003. '*Society must be defended': lectures at the Collège de France, 1975–1976*. London: Picador.

Foucault, Michel, 2005. *The hermeneutics of the subject: lectures at the Collège de France, 1981–1982*. London: Picador.

Foucault, Michel, 2008. *The birth of biopolitics: lectures at the Collège de France. 1978–1979*, London: Picador.

Foucault, Michel, 2010. *The government of self and others: lectures at the Collège de France, 1982–1983*. London: Picador.

Foucault, Michel, 2011. *The courage of truth: lectures at the Collège de France, 1983–1984*. London: Picador.

Friedman, Sara L., 2005. The intimacy of state power: marriage, liberation, and socialist subjects in southeastern China. *American Ethnologist, 32*(2), pp.312–27. doi:10.1525/ae.2005.32.2.312

Fritsch, Kelly, 2010. Desiring disability differently: neoliberalism, heterotopic imagination and intra-corporeal reconfigurations. *Foucault Studies*, [e-journal] 19, pp.43–66. Available at: http://rauli.cbs.dk/index.php/foucault-studies/article/view/4824 [Accessed 8 November 2016].

Gilmore, Leigh, 1994. *Autobiographics: a feminist theory of women's self-representation*. Ithaca: Cornell University Press.

Gökarıksel, Banu, 2012. The intimate politics of secularism and the headscarf: the mall, the neighborhood, and the public square in Istanbul. *Gender, Place and Culture*, 19(1), pp.1–20. doi:10.1080/0966369X.2011.633428

Haraway, Donna, 1988. Situated knowledges: the science question in feminism and the privilege of the partial perspective. *Feminist Studies,* 14(3), pp.575–99.

Hassenzahl, Marc, Heidecker, Stephanie, Eckoldt, Kai, Diefenbach, Sarah and Hillmann, Uwe, 2012. All you need is love: current strategies of mediating intimate relationships through technology. *ACM Transactions on Computer-Human Interaction,* [e-journal] 19(4). doi:10.1145/2395131.2395137

Hayes-Conroy, Allison and Hayes-Conroy, Jessica, 2008. Taking back taste: feminism, food and visceral politics. *Gender, Place and Culture*, 15(5), pp.461–73. doi:10.1080/09663690802300803

Hayes-Conroy, Allison and Martin, Deborah G., 2010. Mobilising bodies: visceral identification in the Slow Food Movement. *Transactions of the Institute of British Geographers* 35(2), pp.269–81. doi:10.1111/j.1475-5661.2009.00374.x

Hekman, Susan, 2010. *The material of knowledge: feminist disclosures*. Bloomington: Indiana University Press.

Hesse-Biber, Sharlene Nagy and Leavy, Patricia, 2006. *Feminist research practice: a primer*. Thousand Oaks, CA: Sage Publications.

Hesse-Biber, Sharlene Nagy and Leavy, Patricia, 2007. *The practice of qualitative research*. Thousand Oaks, CA: Sage Publications.

Hiemstra, Nancy, (in press). Periscoping as a feminist methodological approach for researching the seemingly hidden. *The Professional Geographer*. Published online 30 August 2016. http://dx.doi.org/10.1080/00330124.2016.1208514 [Accessed 8 November 2016].

Jamieson, Lynn, 1989. *Intimacy: personal relationships in modern societies*. Cambridge, UK: Polity Press.

Jamieson, Lynn, 2011. Intimacy as a concept: explaining social change in the context of globalisation or another form of ethnocentricism? *Sociological Research Online*, 16(4), [e-journal] 15. Available at: http://www.socresonline.org.uk/16/4/15.html [Accessed 11 July 2016].

Jamieson, Lynn, 2013. Personal relationships, intimacy and the self in a mediated and global digital age. In: Kate Orton-Johnson and Nick Prior, eds., *Digital sociology*. London: Palgrave Macmillan. pp.13–33.

Jankowiak, William R., 2013. *Intimacies: love and sex across cultures*. New York: Columbia University Press.

Jayne, Mark, Valentine, Gill and Holloway, Sarah L., 2010. Emotional, embodied and affective geographies of alcohol, drinking and drunkenness. *Transactions of the Institute of British Geographers*, 35(4), pp.540–54. doi:10.1111/j.1475-5661.2010.00401.x

Katz, Cindi, 2001. Vagabond capitalism and the necessity of social reproduction. *Antipode*, 33(4), pp.709–28. doi:10.1111/1467-8330.00207

Katz, Cindi, 2008. Bad elements: Katrina and the scoured landscape of social reproduction. *Gender, Place and Culture*, 15(1), pp.15–29. doi:10.1080/09663690701817485

Kaufmann, Jodi, 2011. An autoethnography of a hacceity: a-wo/man-to-eat-androgen. *Cultural Studies ⇔ Critical Methodologies*, 11(1), pp.38–46. doi:10.1177/15327 08610386920

Kern, Leslie, Hawkins, Roberta, Falconer Al-Hindi, Karen, and Moss, Pamela, 2014. A collective biography of joy in academic practice. *Social and Cultural Geography*, 15(7), pp.834–51. doi:10.1080/14649365.2014.929729.

Kivits, Joëlle, 2009. Everyday health and the internet: a mediated health perspective on health information seeking. *Sociology of Health & Illness*, 31(5), pp.673–87. doi:10.1111/j.1467-9566.2008.01153.x

Kobayashi, Audrey and Peake, Linda, 2008. Racism in place: another look at shock, horror, and racialization. In: Pamela Moss and Karen Falconer Al-Hindi, eds., *Feminist geographies: rethinking place, space and knowledges*. Lanham, MA: Rowman and Littlefield. pp.171–8.

Kobayashi, Audrey, 2009. Situated knowledge, reflexivity. In: Rob Kitchin and Nigel Thrift, eds., *International Encyclopedia of Human Geography*. London: Elsevier. pp.138–43.

Kohl, Ellen and McCutcheon, Priscilla, 2015. Kitchen table reflexivity: negotiating positionality through everyday talk. *Gender, Place and Culture*, 22(6), pp.747–63. doi:10. 1080/0966369X.2014.958063

Laliberté, Nicole and Schurr, Carolin, 2016. Introduction: the stickiness of emotions in the field. *Gender, Place and Culture* 23(1), pp.72–8. doi:10.1080/0966369X.2014.992117

Lees, Loretta and Baxter, Richard, 2011. A 'building event' of fear: thinking though the geography of architecture. *Social and Cultural Geography*, 12(2), pp.107–22. doi:10.1 080/14649365.2011.545138

Lejeune, Phillippe, 2009. *On diary*. Edited by Popkin, Jeremy and Rak, Julie. Translated from French by Durnin, Kathy. Honolulu: University of Hawai'i Press.

Letherby, Gayle, 2003. *Feminist research in theory and practice*. London: Open University Press.

Linke, Gabreielle, 2011. The public, the private, and the intimate: Richard Sennett's and Lauren Berlant's cultural criticism in dialogue. *Biography*, 34(1), pp.11–24. Available at: http://www.jstor.org/stable/23541175 [Accessed 8 November 2016].

Longhurst, Robyn, 1994. The geography closest in – the body . . . the politics of pregnability. *Australian Geographical Studies*, 32(2), pp.214–23. doi:10.1111/j.1467-8470.1994. tb00672.x

Longhurst, R. 2012. Becoming smaller: autobiographical spaces of weight loss. *Antipode*, 44(3), pp.871–88. doi:10.1111/j.1467-8330.2011.00895.x

Massey, Doreen, 1994. *Space, Place, and Gender*. Minneapolis: University of Minnesota Press.

McDowell, Linda, 1999. *Gender, identity and place*. London: Polity Press.

McHugh, Susan, 2011. *Animal stories: narrating across species lines*. Minneapolis: University of Minnesota Press.

Miller, Nancy K., 1990. *Getting personal: feminist occasions and other autobiographical acts*. New York: Routledge.

Miller, Nancy K., 2000. Reading spaces: 'but enough about me, what do you think of my memoir?' *The Yale Journal of Criticism*, 13(2), pp.422–36. doi:10.1353/yale.2000.0023

Mills, Sarah, 1997. *Discourse*. London: Routledge.

Milne, Heather, 2014. Dearly beloveds: the politics of intimacy in Juliana Spahr's this connection of everyone with lungs. *Mosaic: A Journal for the Interdisciplinary Study of Literature*, 47(2), pp.203–18.

Morrison, Carey-Ann, Johnston, Lynda and Longhurst, Robyn, 2013. Critical geographies of love as spatial, relational and political. *Progress in Human Geography*, 37(4), pp.505–521. doi:10.1177/0309132512462513

Moss, Pamela, 1993. Introductory comments. Feminism as method. *Canadian Geographer*, 37(1), pp.48–49. doi:10.1111/j.1541-0064.1993.tb01540.x

Moss, Pamela, ed., 2001a. *Placing autobiography in geography*, Syracuse, NY: Syracuse University Press.

Moss, Pamela, 2001b. Writing one's life. In: Pamela Moss, ed., *Placing autobiography in geography*, Syracuse, NY: Syracuse University Press. pp.1–22.

Moss, Pamela, ed., 2002. *Feminist geography in practice: research and methods*. London: Blackwell.

Moss, Pamela, 2005. A bodily notion of research: power, difference and specificity in feminist methodology. In: Lise Nelson and Joni Seager, eds., *A Companion to Feminist Geography*. London: Blackwell. pp.41–59.

Moss, Pamela, 2013. Becoming-undisciplined through my foray into disability studies, *Disability Studies Quarterly*, [e-journal] 33(2), n.p. Available at: http://dsq-sds.org/article/view/3712/3232 [Accessed 12 July 2016].

Moss, Pamela, 2014a. Shifting from nervous to normal through love machines: battle exhaustion, military psychiatrists and emotionally traumatized soldiers in World War II. *Emotion, Space and Society*, 10(1), pp.63–70. doi:10.1016/j.emospa.2013.04.001

Moss, Pamela, 2014b. Some rhizomatic recollections of a feminist geographer: working toward an affirmative politics. *Gender, Place and Culture*, 21(7), pp.803–12. doi:10.1 080/0966369X.2014.939159

Moss, Pamela and Falconer Al-Hindi, Karen, eds., 2008. *Feminisms in geography: rethinking space, place and knowledges*. Lanham, MD: Rowman and Littlefield

Moss, Pamela and Falconer Al-Hindi, Karen, 2014. Feminisms in action: who we are, what we do and how we do it: an introduction to a *Gender, Place and Culture* reader. *Gender, Place and Culture*, [online]. Available at: http://www.tandf.co.uk/journals/pdf/papers/CGPC-Feminisms-in-Action.pdf [Accessed 30 June 2016].

Moss, Pamela and Prince, Michael, J., 2014. *Weary warriors: power, knowledge, and the invisible wounds of soldiers*. New York: Berghahn Books.

Mountz, Alison and Hyndman, Jennifer, 2006. Feminist approaches to the global intimate. *Women's Studies Quarterly*, 34(1/2), pp.446–63.

Mountz, Alison, Bonds, Anne, Mansfield, Becky, Loyd, Jenna, Hyndman, Jennifer, Walton-Roberts, Margaret, Basu, Ranu, Whitson, Risa, Hawkins, Roberta, Hamilton, Trina and Curran, Winnifred, 2015. For slow scholarship: a feminist politics of resistance through collective action in the neoliberal university. *ACME: An International E-Journal for Critical Geographies*, [e-journal] 14(4), pp.1235–59. Available at: http://ojs.unbc.ca/index.php/acme/article/view/1058 [Accessed 11 July 2016].

Norwood, Vera and Monk, Janice, eds, 1987. *The Desert is No Lady: Southwestern Landscapes in Women's Writing and Art*. Tucson: University of Arizona Press.

Nxumalo, Fikile, 2016. Towards 'refiguring presences' as an anti-colonial orientation to research in early childhood studies. *International Journal of Qualitative Studies in Education*, [e-journal] 29(5). http://dx.doi.org/10.1080/09518398.2016.1139212 [Accessed 17 November 2016].

Ojeda, Diana, 2013. War and tourism: the banal geographies of security in Colombia's 'retaking.' *Geopolitics*, 18(4), pp.759–78. doi:10.1080/14650045.2013.780037

Oswin, Natalie and Olund, Eric, 2010. Governing intimacy. *Environment and Planning D: Society and Space*, 28(1), pp.60–7. doi:10.1068/d2801ed

Pacini-Ketchabaw, Veronica, Kummen, Kathleen and Thompson, Deborah, 2010. Becoming intimate with developmental knowledge: pedagogical explorations with collective biography. *The Alberta Journal of Educational Research*, 56(30), pp.335–54. Available at: http://hdl.handle.net/10515/sy5qr4nx7 [Accessed 17 November 2016].

Pain, Rachel, 2009. Globalized fear? Towards an emotional geopolitics. *Progress in Human Geography*, 33(4), pp.466–86. doi:10.1177/0309132508104994

Pain, Rachel, 2014. Gendered violence: rotating intimacy. *Area*, 46(4), pp.351–3. doi:10.1111/area.12138_4

Pain, Rachel and Staeheli, Lynn, 2014. Introduction: intimacy-geopolitics and violence. Area 46, pp.344–7. doi:10.1111/area.12138

Pansoneau-Conway, Sandra, Bolen, Derek M., Toyosaki, Satoshi, Rudick, C. Kyle and Bolen, Erin K., 2014. *Cultural Studies ⟷ Critical Methodologies*, 14(4), pp.312–23. doi:10.1177/1532708614530302

Peterson, V. Spike, 2016. Towards queering the globally intimate. *Political Geography*, [pre-publication view]. Available at: doi:10.1016/j.polgeo.2016.01.001 [Accessed 11 July 2016].

Pratt, Geraldine and Rosner, Victoria, eds., 2012. *The global and the intimate: feminism in our time*. New York: Columbia University Press.

Price, Patricia, 2014. Culture. In: Roger Lee, Noel Castree, Rob Kitchin, Victoria Lawson, Anssi Paasi, Chris Philo, Sarah Radcliffe, Susan M. Roberts and Charles W.J. Withers, eds., *The Sage Handbook of Human Geography*. London: Sage Publications, pp.505–21. doi:10.4135/9781446247617.n23

Professional Geographer, 1994. Women in the field: critical feminist methodologies and theoretical perspectives. 47(1), pp.54–102.

Professional Geographer, 1995. Should women count? The role of quantitative methodology in feminist geographic research. 48(4), pp.426–66.

Rak, Julie, 2009. Dialogue with the future: Philippe Lejeune's method and theory of diary. In: Jeremy Popkin and Julie Rak, eds., *On diary*. Translated from French by Durnin, Kathy. Honolulu: University of Hawai'i Press. pp.16–26.

Ramazanoğlu, Caroline with Holland, Janet, 2002. *Feminist methodology: challenges and choices*. Thousand Oaks, CA: Sage Publications.

Rojas Gaviria, Pilar, 2016. Oneself for another: the construction of intimacy in a world of strangers. *Journal of Business Research*, [e-journal] 69(1), pp.83–93. Available at: http://dx.doi.org/10.1016/j.jbusres.2015.07.023 [Accessed 8 November 2016].

Schultz-Krohn, Winifred, 2004. The meaning of family routines in a homeless shelter. *The American Journal of Occupational Therapy*, 58(5), pp.531–42. doi:10.5014/ajot.58.5.531

Seol, Kwangsoo, Kim, Jeong-Dong and Baik, Doo-Kwan, 2016. Common neighbor similarity-based approach to support intimacy measurement in social networks. *Journal of Information Science*, 42(2), pp.128–37. doi:10.1177/1532708614530302

Sharp, Joanne and Dowler, Lorraine, 2011. Framing the field. In: Vincent J. Del Casino, Jr., Mary Thomas, Paul Cloke and Ruth Panelli, eds., *A companion to social geography*. London Wiley-Blackwell, pp.146–60.

Simonsen, Kirsten, 2007. Practice, spatiality and embodied emotions: an outline of a geography of practice. *Human Affairs*, 17, pp.168–81. doi:10.2478/v10023-007-0015-8

Slocum, Rachel, 2008. Thinking race through corporeal feminist theory: divisions and intimacies at the Minneapolis Farmers' Market. *Social and Cultural Geography*, 9(8), pp.849–69. doi:10.1080/14649360802441465

Smith, Sara, 2012. Intimate geopolitics: religion, marriage, and reproductive bodies in Leh, Ladakh. *Annals of the Association of American Geographers*, 102(6), pp.1511–28. doi:10.1080/00045608.2012.660391

Sprague, Joey, 2005. *Feminist methodologies for critical researchers: bridging differences*. Lanham, MA: Altamira Press.

Stokoe, Elizabeth, 2006. Public intimacy in neighbor relationships and complaints. *Sociological Research Online*, [e-journal] 11(3). Available at: http://www.socres online.org.uk/11/3/stokoe.html [Accessed 12 July 2016].

Tamas, Sophie, 2014. My imaginary friend: writing, community, and responsibility. *Cultural Studies ⇔ Critical Methodologies*, 14(4), pp.369–73. doi:10.1177/1532708614530308

Taylor, Chloe, 2009. *The culture of confession from Augustine to Foucault: a genealogy of the 'confessing animal'*. New York: Routledge.

Thien, Deborah, 2007. Intimate distances: considering questions of 'us'. In: Joyce Davidson, Liz Bondi and Mick Smith, eds., *Emotional Geographies*. London: Routledge. pp.191–204.

Vacchelli, Elena, 2011. Geographies of subjectivity: locating feminist political subjects in Milan. *Gender, Place and Culture*, 18(6), pp.768–85. doi:10.1080/0966369X.2011.617916

Valentine, Gill, 2008. The ties that bind: towards geographies of intimacy. *Geography Compass* 2, pp.2,097–110. doi:10.1111/j.1749-8198.2008.00158.x

Wilkie, Rhoda, 2015. Multispecies scholarship and encounters: changing assumptions at the human-animal nexus. *Sociology*, 49(2), pp.323–39. doi:10.1177/0038038513490356

Women and Geography Study Group [WGSG], 1984. *Geography and gender: an introduction to feminist geography*. London: Hutchinson.

Women and Geography Study Group [WGSG], 1997[2013]. *Feminist geographies: explorations in diversity and difference*. Harlow, England: Longman.

Wright, Melissa W., 2008. Gender and geography: knowledge and activism across the intimately global. *Progress in Human Geography*, 33, pp.379–86. doi:10.1177/030913 2508090981

Wyatt, Jonathan, Gale, Ken, Gannon, Susanne and Davies, Bronwyn, 2010. Deleuzian thought and collaborative writing: a play in four acts. *Qualitative Inquiry*, 10(9), pp.730–41. doi:10.1177/1077800410374299

Zelinsky, Wilbur, Monk, Janice and Hanson, Susan, 1982. Women and geography: a review and prospectus. *Progress in Human Geography*, 6(3), pp.317–66. doi:10.1177/0309132 58200600301

Part I
Methodological challenges

2 An uncomfortable position

Making sense of field encounters through intimate reflections

Maral Sotoudehnia

Foreword

In what follows, I unpack what England (1994, p.87) calls 'intensely personal' writing from my journal that situates my position in the 'field as well as the final text'. Intimate reflections can enable geographers to *act* reflexively and go beyond mere navel gazing exercises that reduce questions of identity to a list of markers (Kobayashi, 2003; Kohl and McCutcheon, 2015). Intimacy can function as an analytical tool that can help frame the uncertain, shifting and contradictory 'layers of sameness and difference' qualitative researchers often encounter (Valentine, 2002, p.122).

An engagement with intimate and autobiographical forms of writing can offer a venue where 'writers, born again in the act of writing, may experiment with reconstructing the various discourses ... in which their subjectivity has been formed' (Gilmore, 1994, p.85). England (1994), for instance, describes research as a necessarily personal experience: professional circumstances are inseparable from epiphanies, encounters and relationships that emerge through the research process. For Gilmore (1994, p.85), autobiography at once constructs and enables access to the very positionings that result from such experiences. Intimate writing encourages researchers to acknowledge how the self is never fully knowable, or what Rose (1997, p.314) calls the 'incoherence of the self' (see also Valentine, 2002, p.125). By laying bare 'the tensions, conflicts and unexpected occurrences' of research through intimate writing, feminist geographers can try to 'decenter our research assumptions, and question the certainties that slip into the way we produce knowledge' (Valentine, 2002, p.126). I therefore engage in what I call intimate reflections, which I define as personal and autobiographical writing. Intimate reflections capture attempts at reflexivity that surround the researcher, research process and some of the conflicting experiences encountered through qualitative fieldwork.

I use intimate reflections as a way to withstand the intellectual squalls of researcher confusion, guilt and worry (Cuomo and Massaro, 2016). Autobiographical impressions from my field journal allow me to confront both the vagaries of field research and the affective capacity of those experiences to generate theoretical insights (Moss, 2001). Through vulnerable and autobiographical scenes that may be represented but never recovered (Gilmore, 1994), I hope to grapple with, however tangentially, some of the personal anxieties I feel towards my 'experiences of hybridity' (Fisher, 2015, p.13) in relation to my Master's research in Dubai, United Arab Emirates.

My research project focused on the neoliberalization of place in Dubai through local place-branding and -naming practices. The study offered an exploratory analysis of the spatial politics of entrepreneurial policies made by various government agencies. Like many graduate students, I was limited by my relative inexperience and knowledge about the research process. As a result, I devoted minimal time and effort to any micro-investigations of my own subject formation or those of my participants. What proved even more cringe-worthy in retrospect was how little I reflected upon my position in a web of power relations and how my actions reproduced and abetted those very same relations. The following excerpt from my journal presents one such instance, where my (willful?) ignorance regarding my positionality becomes apparent.

A summary offence in the field

22 May 2012

My jugular syncopates blood as the policeman examines the contents of my bag. 'They're just notes, officer', I say, anxious to get my stuff back. I can hear the quickening squelch of veins and aorta struggling to circulate my blood. Is that normal? It's the same sound babies make when they start eating solid foods: wet, viscous, thick. Adrenaline and cortisol drip out of me, staining my clothes, gluing my always frizzy hair to my face. I am leaky like a faucet and sweaty like a graduate student about to get arrested in a foreign country. The officer asks me why I 'need to study Dubai' as he pores over my research materials. 'Because it's interesting!' I offer, hoping to charm my way out of this one. Fingers crossed. A sharp pain picks at the stem of my brain and I swallow hard, trying to suppress any fear. Desert hues darken as the rods in my eyes adjust to the setting sun. My mouth is dry. 'Don't faint', I tell myself, as the cop deletes all of my pictures and wipes my voice-recorder clean. Fuck. That had everything and my visa runs out in less than two weeks. Even if I could hear what he's saying over the sounds of my adrenal glands losing their shit, I'm not sure I'd understand anything. Everything appears over-pixelated, grainy. At least he's not going to process me at the station. Just get rid of my project. So long as I can get out of here and not go to jail, it'll be ok. The sound of the Maghrib echoes around us as the officer asks me what a young Iranian woman like me is doing alone and using transit? His voice jumps an octave and becomes mellifluous as he tells me I am beautiful in Farsi worse than mine. Do I have any dinner plans and do I have a boyfriend? Am I married, he wants to know. I lie and say yes, and he bids me farewell. I pick up my bag and try to see what files are left, if any.

In May of 2012, I travelled to Dubai to conduct research for my MA project. I wanted nothing more than to do right by my participants, my committee and the research itself. I cared so much about what my interviewees and supervisors would think of me that I barely considered the impact of the final project itself. Encounters like the one described above, though, tapped directly into a deep well of fear. While my institution's ethics review board cleared me to conduct fieldwork, the ethics of my research and my own were often blurry. Neither my institution nor I anticipated the complex cultural and political negotiations I would have to broker as a single, Persian-Canadian woman conducting qualitative research in an Arab Gulf state. My own ethics, guided by a lifelong fear of inconveniencing, stressing or hurting anything or anyone, were continuously tested. I became aware of the possible stress, inconvenience and hurt the mere notion of questioning the state in Dubai could cause for all involved. What was even more troublesome, I later realized, was how many of my contacts felt obliged to help out a keen student, fellow Iranian, or a Canadian (the descriptors participants used to identify me).

Even though I was an outsider in Dubai, most people I encountered assumed I was a resident. I'm an ethnic blend of Persian, Azerbaijani and countless social groups that once travelled the road to Samarkand, which means that, phenotypically speaking, I *look* like I could belong to Dubai's large Persian Diaspora. But I grew up in Canada. I nonetheless wove in and out of my Persian and Canadian selves to suit my research interests, sometimes purposefully, sometimes unknowingly. After nearly getting arrested for taking pictures of Dubai's new Metro system, I paused to reflect. At the time, I was too rattled to recognize that I likely got away with whatever summary offence I unwittingly committed because I played up my Persian identity. Unpleasant as it was, the encounter left me feeling anything but reflexive about my position, how I practised my own ethics and the ethics of the project itself. In addition to feeling anxiety over the possibility of any future travels in the Emirate and whether or not my encounters with the police would lead to any actionable consequences, I had to find a way to complete the project despite the lost data, my growing (irrational?) fear of the state, and overall dread.

One way to cope with and draw wider insights from any revelations about my hybridity, encounters with an unassuming but powerful legal apparatus or how I was necessarily part of my research, was to write them in the most intimate way and in the most intimate place: my journal. Upon my return to Canada I faced an awkward roadblock to completing my thesis: how could I address my shifting position in the field (and my manipulation of it) and present any results without explicitly attending to my failures to locate myself (honestly) as a researcher amidst the project itself?

Following months of insomnia, I decided to insert intimate reflections into the thesis in hopes of exposing myself and some of the contradictions of mediating my identity in the field. What I presented in these vignettes was based on my impressions of Dubai. The excerpts often involved descriptions of scenes that somehow captured the betweenness of the research process or personal revelations about my position as an individual and (very) junior researcher. The encounters I detailed were personal, but they also illuminated aspects of the research I struggled to articulate, make sense of or reconcile during my field season. These intimate reflections helped me acknowledge and share some of the more troubling experiences I had.

What I had initially discarded as cathartic ramblings while I was in the field became analytical threads I used to hem any incongruities frustrating the completion of my project: my identity in relation to my participants, the research itself or how vulnerable I felt throughout the process. These musings, part autobiographical, part stream of consciousness, offered an avenue by which I could attempt at the very least to communicate my growing awareness of the partiality, ambiguity and vulnerability of what Katz (1994, p.67) calls the 'stories and the artifice of the boundaries drawn in order to tell them' encountered through fieldwork.

Intimate writing + self-reflexivity = intimate reflections

Feminist geographers continue to discuss the importance of self-reflexivity to the research process (Moss, 2001, 2002; Valentine, 2002; Proudfoot, 2015). Through the works of McDowell (1992) and others (e.g. Nast, 1994; Rose, 1997; Moss, 2002), budding feminist geographers are exposed to words like 'self-reflexivity' and 'positionality' across graduate seminars. Both words held collective importance for my cohort: we repeated them like mantras and relied on their incantation to protect us from any unanticipated pitfalls of qualitative research. At the time, we remained ignorant of the cautions and perils of reducing questions of positionality to a mere checklist of identity markers.

I embarked upon my fieldwork with a kit of conceptual and methodological tools, all of which I assumed would protect me, through simple invocation and application, from things going wrong: exploiting my participants, misrepresenting knowledge collected or partaking in any dubious activity as a researcher. As I began data collection, though, my research plan dissolved into an incomprehensible mess. Nothing was rolling out according to plan: getting participants who *did* want to talk to me (mostly out of pity) or sign consent forms was a near impossible task and there were numerous reminders of the ethnic and socio-economic differences that simultaneously connected and separated us from each other. Most of my participants were CEOs, executives of large or multinational corporations, or senior members of Dubai's government. Everywhere I went, participants and people I encountered would identify me based on my complexion or name, and many would label my identity and class accordingly. Who was I to correct them? I operated in a space of betweenness, bound by my perplexing economic and social positions, but also by my ethnic hybridity: atheist, unmarried, Iranian-Canadian woman, poor, student. By contrast, my participants consisted primarily of extremely wealthy ex-pats (many of them Persian) who often looked pitifully upon my oversized blazers, candy bar phone, metrocard and lifestyle.

Falconer Al-Hindi and Kawabata (2002) discuss the discomforts many feminist geographers encounter through 'more fully self-reflexive research', noting that:

> writing about research . . . requires that the researcher identify and locate herself, not just in the research, but also in the writing. She must be willing to re-live the discomforting experiences to look awkward and feel ill at ease. She must commit to paper and thus to the scrutiny of peers and other that which she might prefer to forget. (p.114)

Regardless of my desire to forget the many mortifying failures I was responsible for throughout my degree, my decision to commit my intimate reflections to my thesis and, further, to an academic publication enabled me to confront the inchoate messiness of my position as student, researcher and first generation Iranian-Canadian woman. This helped me, as Falconer Al-Hindi and Kawabata (2002, p.106) explain, 'fill in not only gaps between interviewer and interviewee, but also . . . the spaces that coalesce around it'.

Writing myself in

Before beginning fieldwork in 2012, I actively constructed a research design that positioned me as a complete outsider towards my area of research, which was located 856 km from my birthplace in Kuwait, and a mere 173 nautical miles from my ancestral homeland of Iran. The irony of that decision became apparent as I deplaned in Dubai and worked my way through customs.

I thought, naively, that I would be perceived as an outsider by participants and that introducing myself as a Canadian woman would help me avoid any complicated emotions or encounters. Without the veneer of a 'simplified positionality' I had worked so hard to create (Kohl and McCutcheon, 2015, p.752), the field morphed into a new, terrifying reality. Despite being a Persian girl as one participant put it, I certainly didn't feel very Persian. I had a Western accent when I spoke Farsi, had never been to Iran, didn't identify with most cultural artifacts and cues and was vegan (read: didn't eat *kabob*). The manifold layers of my subject formation were getting too heavy for me to bear without succumbing to the weight of expectation. My journal offered a safe place to store my fears: the deep discomfort I felt trying to fit the mould of a Persian girl, Canadian woman, student or any other version of myself I stumbled across.

4 May 2012

'Flight crew: prepare for landing.' I jolt from a dreamless sleep, my joints aching an ache of airplane seats. Over 24 hours in transit and I am groggy, stiff and become aware of the cabin's recycled air. I want nothing more than to get off this plane. I anticipate the satisfaction that will come when I can finally unfurl my spine, one vertebra at a time. I want so badly to wash the tired off and down an unfamiliar drain. I look across my row and out the window, in hopes of catching a glimpse of Dubai's skyline from above. This is not the Dubai I visited three years earlier, but a different, more developed city. Roads lattice across the landscape and cars travel on roads, organised, like ants. A flight attendant interrupts my thoughts: 'Ma'am? Can you please return your chair to the upright position? We're about to land.' 'Sure', I say, as I release my seat, think about the month ahead and how this city doesn't belong in the Middle East I know, the Middle East I have been taught to know, the Middle East of the West. This city exists in the mythopoeia conjured by grandparents, displaced and forlorn, whose tales hoped to convince me that the Middle East is not inferior to the West, not a place to be forgotten. Sun-bleached photos of family members long past come to mind, framed by modern-looking cities, smiles on their faces.

As I queue behind an endless line of migrant workers at three in the morning, I am bemused by my intake officer, who asks me why an Iranian girl (travelling with a Canadian passport – the only passport I have ever held) is travelling alone and uncovered (no rousari, or headscarf). Following a few moments of uncomfortable silence, the officer waves me through. I am immediately accosted by a group of young men from Tehran who ask in Farsi if I know where to get a SIM card. I respond in my anglicised Farsi to a collective cackling, which is followed by requests for me to repeat various phrases in Farsi for comedic effect. 'But you look so Persian' one of the men explains. It dawns on me that I may have some trouble passing as a Westerner and/or Iranian in Dubai. My identity, however tenuously constructed over the last 27 years, slips away as I snake through the gold-plated bustle of DBX.

Afterword

Intimate reflections can lead feminist geographers to 'truly *act* reflexively' (Proudfoot, 2015, p.1148, original emphasis). Following Proudfoot (2015, p.1148), I am prudent to consider how, through intimate writing, 'the unconscious affects our research as much as identity and privilege do'. Mindful that self-reflexive practices can ensnare researchers in self-indulgence or 'confessional modes of self-narration' (Graeber, 2014, p.82; cf. Proudfoot, 2015, p.1148), intimate reflections on the research experience can nonetheless enable the researcher to gaze at oneself in order to contextualize our roles and practices as facilitators of knowledge collection, creation and communication (Falconer Al-Hindi and Kawabata, 2002). I used personal reflections as a way to face ethical discomforts surrounding my role as both researcher and an instrument of the research and draw attention to the shifting and contradictory positions I operate in and through.

Intimate writing provides an avenue by which to access and present how anxiety and vulnerability collide with the ways by which we locate ourselves in the research process. They can help elucidate, as McKay (2002, p.197) notes, 'how self-understanding changes' as a result of research experiences. Intimate writing as text, data and method can also offer an outlet to engage in meaningful methodological discussions about the common 'worry' and 'awfulness' many of us encounter when conducting research in a specific context (Cuomo and Massaro, 2016, p.94). Including personal reflections enabled me to complete the research without omitting my ineluctable role in shaping and interpreting every experience before, during and after my fieldwork. Intimate reflections helped me come to terms with the 'unique ensemble of contradictory and shifting subjectivities' that I am (Gibson-Graham, 1994, p.219). It was an extremely uncomfortable experience, but as Falconer Al-Hindi and Kawabata (2002, p.114) admit, 'the pay . . . will be worth it'.

Works cited

Cuomo, Dana and Massaro, Vanessa. A., 2016. Boundary-making feminist research: new methodologies for 'intimate insiders.' *Gender, Place and Culture*, 23(1), pp.94–106. doi:10.1080/0966369X.2014.939157.

England, Kim, 1994. Getting personal: reflexivity, positionality, and feminist research. *Professional Geographer*, 1, pp.80–9.

Falconer Al-Hindi, Karen and Kawabata, Hope, 2002. Toward a more fully reflexive feminist geography. In: Pamela Moss, ed., 2002. *Feminist geography in practice: research and methods*. Oxford: Blackwell. pp.103–15.

Fisher, Karen T., 2015. Positionality, subjectivity, and race in transnational and transcultural geographical research. *Gender, Place and Culture*, 22(4), pp.456–73. doi:10.108 0/0966369X.2013.879097.

Gibson-Graham, J. K., 1994. 'Stuffed if I know!' Reflections on post-modern feminist social research. *Gender, Place and Culture*, 1(1), pp.205–24.

Gilmore, Leigh, 1994. *Autobiographics*. Ithaca: Cornell University Press.

Graeber, David, 2014. Anthropology and the rise of the professional-managerial class. *HAU: Journal of Ethnographic Theory*, 4(3), pp.73–88. doi: 10.14318/hau4.3.007.

Katz, Cindi, 1994. Playing the field: questions of fieldwork in geography. *Professional Geographer*, 46(1), pp.67–72.

Kobayashi, Audrey. 2003. GPC ten years on: is self-reflexivity enough? *Gender, Place and Culture*, 10(4), pp.345–49. doi:10.1080/0966369032000153313.

Kohl, Ellen and McCutcheon, Priscilla, 2015. Kitchen table reflexivity: negotiating positionality through everyday talk. *Gender, Place and Culture*, 22(6), pp.747–63. doi:10. 1080/0966369X.2014.958063.

McDowell, Linda, 1992. Doing gender: feminism, feminists and research methods in human geography. *Transactions of the Institute of British Geographers,* 17(4), pp.399–416.

McKay, Deirdre, 2002. Negotiating positionings: exchanging lifestories in research interviews. In: Pamela Moss, ed., 2002. *Feminist geography in practice: research and methods*. Oxford: Blackwell. pp.187–99.

Moss, Pamela, ed., 2001. *Placing autobiography in geography.* Syracuse: Syracuse University Press.

Moss, Pamela, ed., 2002. *Feminist geography in practice*. Oxford: Blackwell.

Nast, Heidi J., 1994. Women in the field: critical feminist methodologies and theoretical perspectives. *Professional Geographer,* 46(1), pp.54–66.

Proudfoot, Jesse, 2015. Anxiety and phantasy in the field: the position of the unconscious in ethnographic research. *Environment and Planning D: Society and Space,* 33(6), pp.1135–52. doi:10.1177/0263775815598156.

Rose, Gillian, 1997. Situating knowledges: positionality, reflexivities and other tactics. *Progress in Human Geography,* 21(3), pp.305–20.

Valentine, Gill, 2002. People like us: negotiating sameness and difference in the research process. In: Pamela Moss, ed., 2002. *Feminist geography in practice: research and methods.* Oxford: Blackwell. pp.116–26.

3 'I'm here, I hate it and I can't cope anymore'

Writing about suicide

Gail Adams-Hutcheson and Robyn Longhurst

Gail's story

On the 6th December 2011 I received a phone call, the impact of which has reverberated ever since. In fact, I still feel a welling up of emotion as I write this chapter. The woman on the other end of the phone told me: 'I'm here [referring to having moved from Christchurch to Hamilton, Aotearoa New Zealand], I hate it and I can't cope anymore.' My diary entry reads:

> I was sitting at my desk in the graduate office when my mobile phone rang. My initial thought was, not *another* potential respondent. My transcribing was already behind schedule. I was right, the Canterbury accent was clear when she asked me about my research on earthquakes. I was puzzled and slightly bemused, almost beginning to be annoyed as Kathryn (a pseudonym – the caller was vigilant not to divulge any personal information) took three attempts to connect with me. When Kathryn did finally connect I was met with silence, then sniffling, followed by halted speech. She blurted out: 'I want to take my own life . . . I, I, I want to die, I've made a big mistake and I don't know how to get out of it.' My heart hammered in my chest, I felt a flush of heat across my face and my head seemed to spin – this was definitely no joke.

No ethics application had prepared me for this moment. I felt naïve and physically sick. Then, as now, I struggle with what to do with this unexpected turn in my data collection.

Although emotions are ubiquitous in research encounters (Bondi, 2005), what do we do with a sense of dread, of horror, of squeamishness, of anger or that which is too raw and too intimate? Feeling rules suggest normative judgements about how one *should* feel in the research encounter (Young and Lee, 1996). When these rules are transgressed, however, it is much harder to engage with the data. Kathryn was careful not to divulge a means of contact (evidenced by making initial contact through a mobile and then switching to a landline), no name, number or hint of her whereabouts. She decided the terms of the call, not me. I note in my diary:

I felt powerless, beyond simply listening, to stop her actions or dictations. Her conversation was fierce and unrepentant about relocating to Waikato. I was embarrassed, caught off guard and scared I may or may not do something to be complicit in her desire for death. But ultimately I felt completely power-less, something which sits so uneasily with a researcher identity.

Reading back over these diary entries five years on, I still feel the pang of surprise and horror as Kathryn blurted out her intention to kill herself. I also feel uneasy and squeamish about writing this chapter, about laying bare my diary reflections for all to read including my co-author, colleague, mentor and supervisor at the time Robyn Longhurst. Several times over the months spent writing this I have wanted to draw back and to submit a different chapter, a more academic and measured piece, but in the final instance a part of me recognized that it is impor-tant to interrogate why this is the case.

Robyn's story

Immediately after the call, Gail phoned me to explain what had happened. She appeared to need reassurance that she had managed the call in the best way pos-sible but this was new territory for me. Despite having two decades of research experience involving many intimate moments with participants (for example, describing their fears about birthing, crying about being fat, and feeling the sting of racism), I had never experienced anything as heart-stopping as someone want-ing to take their own life. Like Gail, I was full of uncertainty, in my case somewhat less about the caller herself since that was out of my control (she'd left no contact details) and more about how best to manage the call from a doctoral student who was calm but clearly very shaken. I was concerned about Gail's emotional well-being. I quickly began to ask myself a barrage of questions: had I adequately trained Gail for this kind of research encounter? Why didn't I see this coming, after all I knew that she was interviewing people who may well be suffering post-traumatic stress? How was I going to best ensure that Gail got adequate follow-up support, care or counselling after such a gut-wrenching call? After ending the call I remember trying to console myself with the thought that if any of my students had to receive such a call then at least I was relieved it was Gail, a mature student with a great deal of life experience who was better equipped to manage the dif-ficult situation than some.

Despite this call having such an immediate and ongoing impact on both me and Gail (we both still wonder to this day whether the caller did at the time, or at some point in the future, take her own life) it is interesting that it made only a brief appearance in Gail's doctoral thesis (500 words out of a total of 100,000, spread over three chapters). She did not, for example, reflect on the experience in any depth in her detailed and reflexive chapter on methods, nor did I advise her to do so. But why?

Often the importance of unexpressed emotion, inexpressible feelings and/ or conflict between thoughts and feelings is overlooked precisely because these things do not easily make sense, nor are they comfortable (Craib, 1995). Gail did not speak about this call in either of two presentations that she gave on emotions in fieldwork. Other participants who appear in the thesis are given a pseudonym, but this participant is referred to simply as 'the caller'. 'The caller' has a kind of absent presence in the thesis – a shadowy figure who lurks but is not fully revealed at any point. Neither Gail nor I, nor a second supervisor, felt comfortable including the experience as data. Nor, however, did we feel comfortable completely ignoring it as though the call had never happened. Some things are difficult to speak about, seemingly horrific and felt so jarring and angular that for years the experiences remain partially or sometimes completely unwritten, unspoken and unintelligible.

Where these stories take us

We began this chapter by describing one particular research encounter – a participant who phoned to say that she planned to take her own life on account of feeling depressed about relocating to Hamilton from post-disaster Christchurch. The use of our first names (Gail and Robyn) to describe this research experience as active participants is deliberate, somewhat unsettling and sits a little awkwardly with us, as our names draw us directly into what has remained largely unspoken in the public domain for some years. In a way the use of we/our *and* Robyn and Gail signals a core argument that we make in this chapter, that is, that it is messy working across multiple subjectivities. Language can stylistically distance authors from their experiences and so we have thought about how we name ourselves and each other both individually and collectively in this chapter.

In the remainder of the chapter we interrogate why it may be that we often do not include stories such as a participant signalling they want to kill themselves in research accounts and what might be at stake in these editing decisions. Even though unused data may stay on the cutting-room floor for some time, the impact of what was said, how it was said and the feelings, moods and residues that research interactions carry with them, at times, continue. Our two overarching research questions are: how and why do we as geographers choose to write specific intimacies, but not others, into our research? And, what might be at stake for participants and researchers in these decisions? First though, in order to provide some context for the call, we describe the PhD project on people relocating to Hamilton and the Waikato region in Aotearoa New Zealand after the Canterbury earthquakes and aftershocks.

In recent years feminist, social and cultural geographers have begun grappling with emotion, affect and non-representational theories in order to facilitate deeper understandings of difference and the multitudinous ways in which the world and all things within it tick. As a result there is now a large and rich literature on the multiplicity of spaces and places that produce and are produced by emotions and affects. Some emotions and affects, however, still appear to be *too intimate* to make their way into geographers' work. These emotions and affects, which are too raw, too close and too personal to air in the public spaces of the academy and beyond, often remain invisible. We see parallels here with geographies of the body. In the 1990s scholars began to address the spatiality of bodies but some aspects of embodiment remained too intimate to discuss, namely their messy materiality. Words such as the body, bodies, embodiment and corporeality began to circulate but other words related to bodies such as 'muscles, mucus, fart, skin, vulva, sweat, urine, hair etc.' (Longhurst, 2000, p.4) were seldom mentioned.

We note a parallel between geographies of emotion and affect and geographies of the body because both are inherently messy yet often represented as 'strangely tidy' (Longhurst, 2000, p.4). This is not something that we are accusing others of, instead this chapter is based on critical *self*-reflection as we consider what might be lost in *not* exploring some of the messy emotional and affective intersections between people and places that are considered *too* intimate to articulate. Both of us have found ourselves *not* writing about research experiences that we feel are too intimate and yet maybe there is a cost to this.

The call in the context of the Canterbury earthquakes

Let us backtrack just for a moment though to explain a little more about the Canterbury earthquakes and to provide a context for our reflections. In 2011, Gail began a PhD that aimed to uncover the stories of 19 families who relocated out of Christchurch in the aftermath of the 2010/2011 earthquakes. Robyn was one of two supervisors for Gail. The investigation involved semi-structured interviews followed by a small focus group. This group in turn resulted in a number of less formal, more spontaneous focus groups held in cafés, parks and in Hamilton's Botanical Gardens. Many of the families interviewed had only lived in the Waikato region for a matter of months. In the main, the interview material centred on participants being able to articulate their earthquake survival stories in a setting beyond Christchurch itself. Without exception the interviews and other exchanges contained heightened emotions linked to surviving earthquakes. The depth of researcher and researched emotional and affective relationships was profound (see Hutcheson, 2013). Respondent descriptions of leaving their friends, family and loved ones as well as the loss of the city they had called home were infused with emotion but were *not* too intimate to write about.

The caller's story, however, about how she hated the Waikato region, about how she felt depressed and regretful about relocating and wanted to take her own life seemed too intimate and too raw to include in any great detail in the thesis, but why? Stylianou (2012) notes: 'The December 23 aftershocks may have contributed to a "significant spike" in suicides in January this year.' The chief coroner outlined that 'there were 81 suicides in Canterbury in the year ending June 30, up from 67 for the 2010–11 year'. This article in *The Press* also notes that the chief coroner says 'it would be "very interesting" to see whether earthquake-hit people who moved to other parts of the country had "affected the numbers in any way".' Clearly there is a need to know more about the effects of the earthquakes and aftershocks in relation to stress, anxiety and depression and yet the suicide call presented many challenges to write into the thesis. It was unknown what the effects might be of writing the encounter into the work. To use the experience as data may be read by some as disrespectful given the graveness of the situation. For those reasons, Gail found it impossible to tell other students, friends and family about receiving the call. Both articulating and capturing the experience in words on the page *felt* wrong and so despite the huge amount of research on geographies of emotion and affect, this story did not find its way into print until now. Like the messy materiality of bodies that Robyn has argued for in other work (Longhurst, 2000), emotions and affects in this instance were tidied up for their brief appearance in the thesis.

Beneath the surface of the neat and tidy

Focusing on how intimate emotions and affects are done within the research context is an important and revealing aspect of feminist, social and cultural geographies that deserves more attention. It is useful to think about the effects of our research not only on our participants but also on our own emotional well-being, and what we write in and what we exclude from our accounts. Some scholars have already begun to pave the way in these directions. For example, more than a decade ago Law (2004) alerted readers to the 'mess of social science research', arguing that often what social scientists attempt to describe is confusing and disordered. Providing simple and clear descriptions is therefore not very effective. A great deal lies beneath the surface of neat and tidy publications. In broaching the intimate, Probyn (2010, p.2) asks researchers to 'replace the empty space of emotionless research procedures with the more interesting one of a being-with: in the space of the intimate'. Valentine's (1998) personal geographies of harassment detail an attack on her identity laying bare a sense of the too intimate. A number of other geographers have also begun the task of investigating these complex, emotional, affectual and intimate participant-researcher relationships (see for example, Bondi, 2005; Longhurst, Ho and Johnston, 2008; Cuomo and Massaro, 2014) but more gritty, messy and intimate work remains to be done in this area.

One line of enquiry might be to question what may well be some the implications of this absence of the too intimate in the work of feminist, social and cultural geographers. It can be difficult to know what to do with highly confrontational research moments. So they may be left to the side. But why might that matter? It is rare that researchers leave the field or data gathering period unscathed or untouched by what respondents have said and how they have said it, and often these poignant encounters can impact deeply on researchers, participants, supervisors and support people. The 'ideology of objectivity' (Haraway, 1988, p.584) that has been so pervasive in scientific discourses for so many years has been nudged aside to make way for some (more acceptable) emotions and affects but not for others.

The question remains, then: is excluding the too intimate problematic in any way, and if so, why? Robyn has argued that invoking bodies in geography that are 'mess- and matter-free' is 'not a harmless omission, rather, it contains a political imperative that helps keep masculinism intact' (Longhurst, 2000, pp.4 and 23).

The messy materiality of bodies is often associated with women, femininity and Otherness. So too are emotions and affects, especially those that are heightened or considered excessive. To address only some emotions and affects (those that are measured, kept in check, everyday and easily managed without fear or embarrassment) and to ignore others that unsettle our sense of self as researchers is another way in which Reason and its long association with masculinism is maintained in the academy. Researchers' lives can be haunted by the ghosts of other stories to be told, other people, places and memories that wait to be enfolded into the everyday mangle of practice. We have focused on one woman's desire to end her own life but there are numerous other stories that researchers might also be reticent to tell. A few examples that spring to mind are sexual desire and/or acts that might take place as part of a research encounter, a researcher prompting a participant to become very upset by a particular line of questioning or a researcher feeling violent towards a participant. Particular topics may also emerge that are thought to be too intimate to address. We have discussed suicide but there are others, for example infanticide, incest, masturbation, necrophilia, bestiality (see Brown and Rasmussen, 2010), and slavery, to name just a few. A number of taboos, then, seem incomprehensible.

Not all stories are easy to tell. Many feelings, emotions and topics remain difficult to speak and write about and yet these are important. Researcher spaces are complex and shifting, they are in excess of expression and never finished. Writing spaces are not static containers waiting to be filled, the material goes well beyond the page. Dewsbury et al. (2002, p.474) argue that 'the world does not add-up, there is always something exceptional, somebody else . . . another example to be given.'

Reflections

We have addressed why the intimate is chosen (or in this case not chosen) for circulation in the public sphere, whether it be for presentation or publication. We identify a parallel between a lack of messy bodies being included in research and a lack of messy and/or raw emotions and affects being included as serious data in our projects. Research participants, not just through their words but also through their bodies, express a range of emotions that give rise to particular affectual research encounters. Sometimes, however, we are not particularly well-prepared for the intensity of these intimate encounters, leaving us feeling uncertain as to how best to write them into our texts and our academic presentations. As a result we often simply leave them out. They do not fit comfortably in our carefully reasoned and rational accounts of research. The academy is no place for heightened or uncontrollable emotion. And yet there is still little investigation that centres on the emotionally charged decisions about what goes into the public sphere and what stays locked up gathering dust until it has reached its best-by date, is shredded, or may come to light some years after the interview. We argue that although geographers over the past decade have begun to write emotions and affects into their texts, some of the more intimate feelings remain closeted. Feminist geographers have long paid attention to power relations in research, reflexivity and situated knowledge production. Some recent examples include Kaspar and Landolt (2016) on flirting, Diprose, Thomas and Rushton (2013) on sexual desire and Smith (2016) on angst. But as yet, there is still a range of other emotions and affects that appear to be deemed too revealing, for example, suicide. It is our hope that we have been able to deepen these discussions by focusing on the too intimate in geography and look forward to other contributions in this area.

Works cited

Bondi, Liz, 2005. The place of emotions in research: from partitioning emotion and reason to the emotional dynamics of research relationships. In: Joyce Davidson, Liz Bondi and Mick Smith, eds., *Emotional geographies*. Aldershot, England: Ashgate, pp.231–46.

Brown, Michael and Rasmussen, Claire, 2010. Bestiality and the queering of the human animal. *Environment and Planning D: Society and Space*, 28, pp.158–77.

Craib, Ian, 1995. Some comments on the sociology of emotions. *Sociology*, 29(1), pp.151–58.

Cuomo, Dana. and Massaro, Vanessa, 2016. Boundary making in feminist research: new methodologies for 'intimate insiders.' *Gender, Place and Culture: A Journal of Feminist Geography*, 23(1), pp.94–106.

Dewsbury, John David, Harrison, Paul, Rose, Mitch, and Wylie, John, 2002. Enacting geographies. *Geoforum,* 33(4), pp.437–40.

Diprose, Gradon, Thomas, Amanda and Rushton, Renee, 2013. Desiring more: complicating understandings of sexuality in research processes. *Area*, 34(3), pp.292–98.

Haraway, Donna, 1988. Situated knowledges: the science question in feminism and the privilege of partial perspective. *Feminist Studies*, 14, pp.575–99.

Hutcheson, Gail, 2013. Methodological reflections on transference and countertransference in geographical research: relocation experiences from post-disaster Christchurch, Aotearoa New Zealand. *Area* 45(4), pp.477–84.

Kaspar, Heidi and Landolt, Sara, 2016. Flirting in the field: shifting positionalities and power relations in innocuous sexualisations of research encounters. *Gender, Place and Culture: A Journal of Feminist Geography*, 23(1), pp.107–19.

Law, John, 2004. *After method: mess in social science research*. London: Routledge.

Longhurst, Robyn, 2000. *Bodies: exploring fluid boundaries*. London: Routledge.

Longhurst, Robyn, Ho, Elsie and Johnston, Lynda, 2008. Using 'the body' as an 'instrument of research': kimch'i and pavlova. *Area*, 40(2), pp.208–17.

Probyn, Elspeth, 2010. Introduction: researching intimate spaces. *Emotion, Space and Society*, 3(1), pp.1–3.

Smith, Sara 2016. Intimacy and angst in the field. *Gender, Place and Culture: A Journal of Feminist Geography*, 23(1), pp.134–46.

Stylianou, Georgina, 2012. Aftershocks linked to suicide spike. *The Press*, [online] Available at: http://www.stuff.co.nz/the-press/news/christchurch-earthquake-2011/7602992/Aftershocks-linked-to-suicide-spike [Accessed 3 September 2015].

Valentine, Gill, 1998. 'Sticks and stones may break my bones': a personal geography of harassment. *Antipode*, 30, pp.305–32.

Young Elizabeth and Lee, Raymond, 1996. Fieldworker feelings as data: 'emotion work' and 'feeling rules' in first person accounts of sociological fieldwork. In: Veronica James and Jonathan, Gabe, eds., *Health and the sociology of emotions*. Oxford: Blackwell, pp.97–113.

4 In the skin

Intimate acts in economic globalization

Maureen Sioh

> Fugue: In music the word refers to a composition in which three or more voices enter imitatively one after another. . . . The first voice to enter carries the principal theme of the fugue which is known as the 'subject'. After this theme has been presented the second voice enters, transposing this subject to the dominant: these dominant entries are called 'answers'. The third voice then enters after the subject, and so it continues: an alternation of subject and answer until all the voices in the fugue have entered.
>
> G.M. Tucker (1983, p.731)

In 2011 I began studying the motivation behind the economic decisions of two government-linked corporations in East Asia. Each has acquired iconic status internationally, yet has generated hostility and anxiety beyond the economic arena. A Government Linked Corporation (GLC) is a company in which the state holds controlling ownership. I use GLC1 and GLC2 to refer to these two corporations: GLC1 is the largest company in the world of its kind in terms of land cultivated with that specific crop and GLC2 is a sovereign wealth fund, capitalized at US$240 billion, a small figure but a significant one given the size of the country it represents and its small global staff of under 500 employees. To put it in economic perspective, GLC2's capitalization is considerably less than the US$2.6 trillion capitalization of the banking conglomerate JPMorgan Chase. Because of the political sensitivity of the subjects and their governments' predilections for litigation, I have chosen to leave the names of the GLCs unspecified. East Asian emerging economies, many of which are also postcolonial, receive about 30 per cent of the inward FDI (Foreign Development Investment) while themselves providing about the same amount of outward FDI. China, Hong Kong (China) and Singapore alone accounted for $420 billion of inward FDI (IMF, 2015; UNCTAD, 2015). These figures are highly volatile given the turmoil in the Chinese stock market in August 2015 followed by the devaluation of the yuan. While these GLCs are spearheading the economic and geopolitical strategies of emerging economies, they also act within the rules of the global financial sector, which they recognize as having been constructed within a cultural and historical matrix.

Using this study as an entry point into some thoughts on intimacy. I engage with the questions of how and what we hear and talk about when we talk about economics. The links between globalization, capitalism and colonialism have been well documented (e.g. Sheppard, et al., 2009; McMichael, 2011), but in my study I was interested in the existential cascade of consequences from these links and how they might influence decisions in global finance. In answering this question, my study became about negotiating trade-offs between gains and losses in financial, cultural and emotional capital: for the interviewees as they made their investment decisions and between the interviewees and myself as I attempted to map the negotiations their professional lives entailed. Our histories – mine and theirs – are connected through colonialism and capitalism so that we borrow each other's stories until it is hard to distinguish origins. It is in this sense that globalization as well as the research process approximates the musical fugue of the epigram – as social beings we build consciously and unconsciously on ideas that swirl around us so that our original ideas return to us reverberated through multiple voices. In this chapter, I want to explore the negotiation between emotional exposure to and the risk of betrayal, and the ethical dilemmas attendant on intimacy, as a precondition to my research process. The world of finance capital becomes a relatively contingent context through which I illustrate my reflections. I engage with three questions. Why does psychoanalysis matter to a study of economic globalization? How, as researchers, do we access the psychocultural mechanisms at work in decision-making? And, what are the burdens of accessing this knowledge? At the heart of all three questions lies the issue of intimacy: its promise as an analytical technique and its danger of moral duplicity. I have written on the first question elsewhere but here I take the opportunity to focus on the second and third questions – the bookends to research that usually get left out of the written narrative.

Embodying the local in the global

I use the term globalization, mindful of the popular connotations that come with it, in the sense of the scale of interaction economically, culturally and politically. My study builds on Appadurai's (1996) conceptualization of the tensions between homogenization and heterogenization in the collision between the local and the global. I question how identities, and thus decisions, are shaped by the disjunctures and contingencies of interactions at the global scale. Yet economic globalization, to take one manifestation of change at the global scale, is felt at the scale of the body. Conversely, if change at the local scale is affected by systemic change at the global scale, global changes are performed and materialized by actual bodies in a physical space. As Appadurai argues, people navigate between their own local experience and graft new ideas from globalization. This is a subtle but important distinction in understanding what practical steps relatively disempowered nation-states can take to protect themselves. The danger lies, not as Pratt and Rosner (2012, pp.2–4) note, in assuming a deterministic asymmetry in power due to binary distinctions, but in forgetting that even within the asymmetry, non-Westerners have agency. Thus, rather than separate spheres, their emotional responses bleed across spaces and social registers. If political and economic decisions are driven by emotional investments, we need to understand what those investments are and how they come about (Sioh, 2010; 2014). The stories we tell form a way into understanding those investments. These stories can only be accessed once a human connection is established.

Intimacy as a practice can help create a safe space in which to access these stories. There is no universal definition of intimacy but the term loosely is taken to encompass the types of personal relations that are subjectively experienced and socially recognized as close, but not necessarily bodily or sexual (Jamieson, 2011, p.8). To be intimate is to discover that you and the other person share a fluency in language that you do not have with others, a fluency that has more to do with the nuances of shared meaning, perhaps through experience, rather than elaborate vocabulary. Jamieson (2011, p.3) suggests that intimacy involves mutual exposure or self-disclosure, giving each other undivided attention, or reciprocating in aid. The last is an assumption in much social science research, especially critical studies: that either the research subjects are members of a subjugated group or the socio-economic peers of the researcher. In my study, this assumption obviously did not hold as my interviewees were members of the elite in their countries. I should note that every one of my 14 interviewees gave me their undivided attention by not taking calls or allowing interruptions during the interviews. But if intimacy suggests something hidden away from the world (Pratt and Rosner, 2012, p.4), then there is always the implied threat of betrayal inherent in exposure.

Our 'horizons of experience' (Rose, 1996, p.2) have greatly expanded through colonialism and globalization. Psychoanalysis, as Kingsbury and Pile (2014, p.5) so aptly put it, opens up 'a way for us to understand the unconscious on the outside; the world's unconscious worlds'. Through its focus on unconscious drives, repression, disavowal, projection and fantasy, psychoanalysis opens up the limits of what is thinkable and challenges the actions that we take for granted. At its most basic, examining our capacity for self-understanding matters because emotions can be manipulated politically (Pile, 2010, p.7). Discrimination due to race, gender and sexual orientation are intimate hurts that translate at the global scale into inequality and war. In my study, I was trying to understand the forms global finance takes when it is felt in the skin.

Within the above framework I pose the question of how the agents of an emerging economy construct the meaning about the country's participation in economic globalization, building on the possibilities and limits posed by their own experience. These subjects of contemporary economic globalization continue to live according to the rules of the game that have been devised elsewhere, yet they also negotiate the tension between acceptance of and resistance to these rules (Sioh, 2010; 2014). My study had to devise a way of broaching and obtaining responses to how my interviewees judged the costs and benefits of the decisions they faced. Traditional psychology assumes the individual is constituted independent of social forces rather than constituted through intrasubjectivity, a view which decontextualizes the individual from politics and history. This separation is important because it elides the ways in which social and political power can precede and constitute an individual. The challenge in applying a psychoanalytical approach to global finance lies in figuring out how to assess when such an approach is merited, and second, how to integrate the theory into the actual research process in such a challenging political environment that restricted the capacity to form trusting relations in a limited time frame.

Background

GLC1 is a publicly traded multinational conglomerate with a workforce of over 100,000 employees established as a European private company in 1910 before being nationalized in 1981. In January 2007 it merged with three other GLCs from the same country. Its directors over the years have tended to be ex-politicians or senior civil servants and its majority shareholder is another GLC. Its investments range across energy, utilities, property and healthcare in 20 countries although its core revenue generator remains agricultural. In the early 2000s, as the national government realigned its globalization strategy to exploit strengths in the agricultural sector, the GLC expanded plantation operations overseas in Asia and Africa. Today its land bank stands at over 800,000 hectares.

GLC2, founded in 1974, is an independently managed commercial investment company with a multinational staff of 490 that invests in financial services, telecommunications, media and technology, transportation and industrials, life sciences, consumer and real estate, as well as energy and resources. Over 70 per cent of its investments are in Asia, a deliberate strategy meant to acknowledge and support the economic aspiration of its neighbours, but it is increasingly expanding its activities in Europe and North America as sites with higher rates of return and safer investments. As with GLC1, its officials are drawn from the ranks of government and the corporate sector.

Intimacy as psychoanalytical methodology

Most psychoanalytical studies in the social sciences rely on public documents or publicly available artifacts such as literature, artwork or pop culture. But the aim of this study was to try to understand the rationale of the decision-makers. While a psychoanalytical approach would acknowledge that what I was told was not necessarily the whole story – indeed, the speakers would not have been conscious of the excess beyond their representations – if we take agency and resistance as actual possibilities, then listening to the voices of the decision-makers is not optional. I have stated that I use the musical fugue as the model for how we build up and upon the knowledge, conscious and unconscious, that underpins our decisions. Knowledge sharing in research, especially when it involves politically sensitive material, requires an intellectual connection *and* an emotional one because the interviewee has to trust that the interviewer understands the stakes involved without articulation. An emotional response is needed to create a context for trust when we don't know each other. The violinist Gidon Kremer describes his performance as an attempt to reach out to a heart in the audience waiting for his sound. In undertaking intimate research, we neither know who that waiting heart will be, nor how our words will be heard. But a fugue can only come into being through that heart. If intimacy as a research practice is oriented towards disclosure and emotional exposure, then the risk of betrayal means that the individual becomes a significant factor in intimacy because her presence signals the boundary and territory of a safe space.

Finding candidates who qualified and were still willing to talk to me proved difficult. Several potential interviewees cited legislation and their governments' known penchant for litigation as reasons for not participating. The information used for my analysis comes from 14 personal interviews. In selecting the candidates for the interview, the only criterion (other than their willingness to speak to me) was that they had to have had a major role in one of the GLCs or a subsidiary's decisions. I define a major role as having a direct say or having an advisory role in projects ranging in worth from tens of millions to billions of dollars. The interviews lasted a minimum of two hours, with the longest lasting seven and a half hours. In accordance with the wishes of the interviewees, I have left out certain parts of the interviews that they deemed too politically sensitive. I also left out those parts that I judged to compromise confidentiality.

I had no working precedent of how to conduct a study lodged in the nexus of finance, postcolonialism and psychoanalysis. In a world replete with studies on money, race and sex, my surprising discovery was how little theories engage with one another. Less surprisingly, I found that research methodologies dealing with interviewing powerful individuals were even less common. My biggest challenge was how to gain the cooperation of interviewees who were powerful both economically and politically, and lived and worked in an environment fraught with the threat of lawsuits and prison sentences. My first and most immediate concern was how to negotiate my degree of personal exposure. While conducting research in a conservative environment facing powerful individuals in a politically tense situation, the only negotiating tool I had was how much I was willing to reveal of myself in return for their trust. The second concern to emerge during the interviews, which became more pronounced during the writing and presentation engagements, was how to approach and convey a story that was very much disavowed in official government narratives. Interestingly, it was easier in the intense interview atmosphere to approach sensitive subjects as the interviewees were much more candid. In most cases, the interviewees brought up sensitive topics without my prompting. And the final negotiation was with my conscience – the extent to which I was willing to use the information I obtained.

Native speaker

Research on globalization, and especially on finance in non-Western societies by Westerners, almost always runs the risk of essentializing its research subjects because of the inability to communicate with any degree of nuance in the absence of shared history and culture. Moreover, while factory workers and peasants are the subjects of ethnographies, there are no critical ethnographies of financial elites, who, if they allow themselves to be interviewed at all, will be interviewed by the business press and will convey pre-packaged statements. I use the term native speaker in the sense of shared cultural and local knowledges rather than a common language. Ethnographic interviews also require that participants establish some measure of trust that their words will not be taken out of context; this trust has to be established between people who have never met each other and, because they are busy, only have a limited time to devote to an interview. Thus, any meaningful study has to be preceded by communication that establishes trust that has nothing to do with the confidentiality disclaimers that researchers are obliged to hand out. Establishing the parameters of trust has to come from both sides. In my study, the most frequently referenced parameter was common acquaintances. One interviewee introduced me to his wife saying that I had been at school with someone they knew. Another interviewee insisted that we were both acquainted with a banker who was a well-known political insider. But less obvious touchstones were equally important; one interviewee talked about the social customs of a local school with which we were both familiar. Like a fugue, the same subject in the interviews would be picked up by each interviewee starting from a different note.

Shared memories also set common ground. For example, one interviewee brought up the memory as a young man of watching the sunrise over a stretch of coastline from a ship knowing that I had often visited that stretch of coastline. Despite coming from different perspectives, the occasion can stand as a metaphor for a common vision. One interview, at the request of the interviewee, was held in a part of town in which people of my ethnic background had traditionally lived. The interactions appeared to be attempts to gauge whether we recognized and were sympathetic to the same cultural signposts before we got down to the main discussions in which the political and economic issues would be foregrounded. These interactions, like a fugue, seemed like questions that called forth an answer from me: could I be trusted? The point here is that shared memories on commonplace topics of food, local beauty spots and school rivalries, even specific old growth trees, established whether I could be trusted to hear *how* they wanted to be heard. The ability to switch between languages helped, particularly when interviewees resorted to kitchen idioms to convey a point. The fact that I could follow them in a language that recalled intimate spheres of childhood and the domestic established an intimate context for the interviews.

Perhaps one incident, more than any other, encapsulates the powerful unspoken assumptions of race, class and gender in establishing a comfortable context for the interviewees. As I entered a taxi on my way to an interview, the driver looked at me in the rearview mirror and before even asking me where I was going, interrogatively commented, 'Brunch?' In a moment, that single word summarized all the assumptions that would be mobilized in the way I would be seen by strangers.

In the skin

The desire by my interviewees to be heard in the right manner by me required assessing a commensurability of values, in which aesthetic judgements or familiarity with local cultural issues became stand-ins for whether the interviewees felt they would be given a sympathetic hearing. In conventional research protocol, this would be evidence of an ideological and, hence, research bias. But my point here is that the search for emotional connection underpins any dialogue between two people to facilitate whether information is heard, not merely exchanged. Every interviewee assumed that I recognized how culture and, by default, culture as race, structured our conversations on the rules of global finance and politics and the ways in which race translated into monetary value. Like a fugue, the theme of race was picked up by each interviewee and returned to me reverberated through their different voices. This recognition was obviously important to the interviewees but rarely addressed in either business or academic literature on financial decision-making. The conversation invariably moved on to race anxiety and a psychology of crisis that framed economic decision-making. For example, commenting on GLC1's investments in a conflict ridden country notorious for human rights violations, one interviewee said that putting aside moral considerations was simply the price nationalities paid for being at the bottom of the international pecking order. Similarly, discussion of infrastructure projects casually assumed a race price tag, an assumption echoed by interviewees from GLC2.

GLC2 has an international reputation for meritocracy, of which its staff and management are justifiably proud. Yet the concept of race, used interchangeably with culture, permeated the interviews, sometimes as the focal point, but more startlingly, in casual asides. The most striking example offered was the case of a Chinese bank's initial public offering (IPO). As a gesture of mutual respect, GLC2 was offered shares in the IPO, which they purchased. But as the Western financial press and investors began to voice concerns about the Chinese bank, GLC2 began to doubt its investment, culminating in the sale of the shares. This gesture not only had the material repercussion of driving down the share value, but also, more significantly in East Asian societies, the gesture was interpreted as disrespectful and questioning the viability of the Chinese bank causing the Chinese government to lose face. The behind-the-scenes diplomatic tension was eventually resolved when GLC2 reinvested in the bank's shares. Yet tensions continued to simmer and the situation was highlighted by my interviewees as an example of an unnecessary difficulty caused by an unwarranted faith in Western judgement. GLC2's actions could be interpreted as a defensive measure to protect its investments. But even within its ranks, the divestment action was interpreted as a simultaneous lack of confidence and passive acceptance of Western judgement. In another example, GLC2's purchase of shares of a beleaguered GLC, much derided by the Western financial press, was described as a justifiable 'message to foreigners' by three interviewees. One interviewee had recently overseen the multibillion dollar sale of technology to a BRIC country and candidly admitted that the deal was sealed because his country was seen as more trustworthy than its Western competitor, which was likely to short-change non-white countries. In this sense, my interviewees, while overtly claiming that meritocracy would win out, were also tacitly acknowledging a power asymmetry that made the above actions perfectly justifiable as defensive gestures. Given their powerful public profiles, discussing the emotional irrational dimension of the decision-making process could only take place if they felt comfortable with me.

Fugue

All my interviewees are extremely successful in the global economy. The experiences of my interviewees contradict Western business models that claim business outcomes are unaffected by national and cultural considerations. They acutely recognize the realities of cultural power in making the rules of the global market and incorporate that recognition in their decision-making even as they simultaneously disavow any influence beyond market discipline. The assumption in the Western business press that these men were merely businessmen or political operatives is unwarranted – to a man, they had a keen awareness of history and cultural politics. What my interviewees wanted to establish was whether I could empathize with their perspective and pick up on the nuances in their financial narratives.

There is always the question of commensurability between what is said by one person and whether it is understood in the same manner by another. Besio (2005, p.323) claims that in trying to find an idiom that each other recognizes, research subjects can become cultural producers of the story of globalization. My interviewees are cultural producers of a different globalization from Westerners, despite shared similarities. As decision-makers in the GLCs, they are resisting in their own fashion on behalf of their states. This is important in the context of what states can do in economic globalization. These men are not members of a conventionally subjugated group yet there is a power differential internationally that they acknowledge and incorporate in their decision-making. Their acknowledgement has material consequences in healthcare, education, and housing, which in these countries are subsidized by state revenues. To ignore this power differential is to participate in a homogenizing discourse that ignores the aspirations of the billions who depend on the ability of their states to provide the basis of realizing their aspirations. The point I want to make is that the production of a discourse is not necessarily a conscious one nor simply an effort at finding a common idiom. Earlier I referred to the metaphor of the musical fugue. But music is both an intellectual and emotional medium of communication. My interviewees expected that the final story I constructed would be built out of the common cultural and historical understanding of what went into the aspirations of these countries that then bled into their decisions they made, which they as agents of the GLCs stood in for their governments. And like a fugue, the common subject of race and culture, taken up by each interviewee, would return to me reverberated through multiple voices but with the original subject intact. They relied on the unspoken and perhaps unconscious assumption that I cared enough to find commensurability in politics and economics given that they now knew we shared memories, local knowledge and the kitchen idiom, which itself harks back to the emotional content of the domestic sphere. That I would write with a listening heart.

But the listening heart can also be a betraying voice. My bodily presence demarcates the boundaries of a safe space for intimacy, yet paradoxically, I become involved in emotional fencing and exposed to the danger of personal questions of identity and family. These questions were inevitable and sometimes essential to establishing trust as some interviewees talked about personal matters. Such a situation led to split-second decisions as I had to calibrate the degree of reciprocity in terms of self-revelation. One situation, the issue of identity, through marital choice and race became uncomfortably testy. The cost of such gestures is high and it is unlikely that I will ever carry out similar research again. It bears remembering that even in this supposedly most rational of topics, financial invest-ment, the interviewees and I made mutual attempts at an emotional calculus that did not always succeed. More to the point, in long and intimate conversations that are one-sided, the speaker can inadvertently reveal sensitive information. All my interviewees were politically sophisticated men who would never reveal proprietary trade information but I wasn't perceived as a business rival nor was I a member of the business press. Candid comments about business practices, especially the explosive ones involving nationality and race in controversial deals were revealed casually. As a critical scholar, I found the material was sometimes exactly what I needed for my study, in which case the decision became mine as to whether to use information that was easily traceable. I have chosen not to use the information even when there was no injunction against using it. Betrayal on my part would have been less to do with financial disadvantage as to do with providing the potential for humiliation, which returns us to the psychoanalyti-cal dimension of the study in the sense of the unspoken and unconscious stakes in economic choices. The writing itself, part of a dialogue between myself and the reader, extends into a negotiation with myself on what it means to write on economic decisions. At its most successful, intimacy as method can be a form of emotional seduction – intimacy's other face to professional success is moral duplicity, a reflection no one would wish for.

Haunted choices

I used this study on the investment decisions of GLCs in East Asia as an entry point to engage with the questions of how and what we hear and talk about when we talk about economics. My study became about the negotiations between the interviewees and myself on how to convey information about finance capital mediated through cultural and emotional capital as I attempted to map the psych-ocultural mechanisms at work in postcolonial international finance. Intimacy as a research practice was central to accessing this knowledge but its gifts would also constitute an ethical burden.

The metaphor of the musical fugue has been my guide throughout this study. We construct stories about our world by building consciously and unconsciously on ideas around us and on each other's hopes and fears – the research may begin with a subject, as in the fugue, but the fugue/research only comes into being when all the voices have entered into the conversation and answered each other. The reader or listener becomes the final voice in the fugue who receives the pre-vious voices as a whole and makes of it what she will. My interviewees expected that the final story I constructed would be built out of the common cultural and historical understanding of what went into the aspirations of these countries that then bled into their decisions. Their trust came through interactions to gauge whether I recognized and was sympathetic to the same cultural signposts before we got down to the interviews on financial investments. Shared memories on commonplace topics framed whether I could be trusted to hear *how* they wanted to be heard: that I would care enough to hear the nuances in superficially cold rational economic narratives.

The result was the unspoken assumption that I got the ways in which race translated into value; indeed, that the value was far beyond monetary. Getting, in this sense, meant being aware of the anxieties inherent in their decisions, but more significantly, it meant a recognition of how much they and their countries have achieved because I recognized the limits with which they began. But if my bodily presence demarcated the boundaries of a safe space for intimacy, the same body now poses the danger of exposure to humiliation for the interviewees who trusted me. As a scholar, success is gauged by access to relevant material but the unspoken caveat in studies on motivation is that the price is often emotional exposure, which opens up anxieties of betrayal – for my interviewees of course, but also for my own self. Betrayal for me would have been the betrayal of the ethical self I like to believe is who I am. Leaving aside the political danger referred to at the start of the chapter, I have chosen to leave out material even when there was no injunction against using it. Betrayal of my interviewees on my part would have been providing the potential for humiliation. When I began this study, I privately thought of it as a love letter to a country. Now I know that like all love letters, mine bears within it the seeds of betrayal. The question is, how far would I go? I have come to accept, not far at all.

Acknowledgements

I wish to thank Pamela Moss and Courtney Donovan for their editorial comments. I am also grateful to one reader who wished to stay anonymous for comments on an early draft. All errors are my sole responsibility.

Works cited

Appadurai, Arjun, 1996. *Modernity at large: cultural dimensions of globalization.* Minneapolis: Minnesota University Press.

Besio, Kathryn, 2005. Telling stories to hear autoethnography: researching women's lives in Northern Pakistan. *Gender, Place and Culture,* 12(3), pp.317–33.

IMF (International Monetary Fund), 2015. *Cross currents.* [online] Washington: Internal Monetary Fund. Available at: http://www.imf.org/external/pubs/ft/weo/2015/update/01/ [Accessed 12 January 2015].

Jamieson, Lynn, 2011. Intimacy as a concept: explaining social change in the context of globalisation or another form of ethnocentricism? *Sociological Research Online,* [online] 16(4). Available at: http://www.socresonline.org.uk/16/4/15.html. [Accessed 15 January 2016].

Kingsbury, Paul, and Pile, Steve, 2014. Introduction: the unconscious, transference, drives, repetition, and other things tied to geography. In: Paul Kingsbury and Steve Pile, eds., *Psychoanalytic geographies*. Surrey: Ashgate, pp.1–38.

McMichael, Philip, 2011. *Development and social change: a global perspective*. Thousand Oaks: Sage.

Pile, Steve. 2010. Emotions and affect in recent human geography. *Transactions of the Institute of British Geographers*, 35(1), pp.5–20. doi:10.1111/j.1475-5661.2009.00368.x.

Pratt, Geraldine, and Rosner, Victoria, 2012. Introduction: the global and the intimate. In: Geraldine Pratt and Victoria Rosner, eds., *The global and the intimate: feminism in our time*. New York: Columbia University Press, pp.1–27.

Rose, Nikolas, 1996. *Inventing ourselves: psychology, power and personhood*. Cambridge: Cambridge University Press.

Sheppard, Eric, Porter, Philip, Faust, David, and Nagar, Richa, 2009. *A world of difference: encountering and contesting development*. New York: Guildford Press.

Sioh, Maureen, 2010. The hollow within: anxiety and performing postcolonial financial policies. *Third World Quarterly*, 31(4), pp.581–98.

Sioh, Maureen, 2014. A small narrow space: postcolonial territorialization and the libidinal economy. In: Paul Kingsbury and Steve Pile, eds., *Psychoanalytical geographies*. London: Ashgate, pp.279–94.

Tucker, G.M., 1983. Fugue. In: Denis Arnold, ed., *The new Oxford companion to music*. Oxford: Oxford University Press, p.731.

UNCTAD (United Nations Conference on Trade and Development), 2015. *World investment report: reforming international investment governance* Geneva: United Nations. [pdf] Available at: http://unctad.org/en/PublicationsLibrary/wir2015_en.pdf [Accessed 8 November 2016].

5 Navigating intimate insider status

Bridging audiences through writing and presenting

Vanessa A. Massaro and Dana Cuomo

Introduction

In spite of the increased attention in geography towards intimacy and intimate field research, we have observed reluctance among scholars – ourselves included – to delve fully into explaining the intimate connections that we share with our research participants. Our concern with uncovering some of the nuances of this kind of intimate research largely stems from our position as intimate insiders in the communities we study. In both our cases, we shared pre-existing and intimate relationships with our research participants. These relationships enabled the level of access obtained in our research projects. Since the completion of our fieldwork, we have struggled with how much to reveal about our intimate insider identities within academic forums. Tensions between the perspectives and experiences of our academic and research communities are at the heart of this struggle.

As intimate insiders, our lives constantly straddle two distinct worlds. We both maintain complex, intimate ties to our research communities beyond our role as researchers. We live and work with our research participants, while maintaining our professional identities as academics. For Vanessa, living in her field site (a low-income, primarily Black neighbourhood, called Grays Ferry, in Philadelphia), means traversing the difficult terrain of poverty, violence and mortality. She often finds herself in the jarring position of transitioning between her field site (and her intimate position as partner, mother and neighbour) and the administrative responsibilities of university life. Similarly, Dana moves between her field site in a police department (where she used to work as a victim advocate for survivors of domestic violence) and an academic community highly critical of policing and the criminal justice system.

We identify strongly with the value systems and perspectives of both our research participants and our academic colleagues, even when those perspectives are at odds. For example, in Grays Ferry, Vanessa's research participants portrayed a glib casualness towards death and violence that made sense given their daily experiences of violence. However, this perspective conflicted with the ideals of human rights and anti-violence upheld in critical circles of the academy. Similarly, in Dana's fieldwork, an aggressive police response and incarceration could play an important role in effectively protecting victims of domestic violence.

Yet in the academy, critical scholars of mass incarceration prioritize abolitionism as a means to challenge the racism and violence of state policing and imprisonment. We both struggled to bridge the competing value systems we faced.

Our dual commitments do not make us unique among academics; indeed, the cornerstone of feminist scholarship concerns the bridging of theory and practice (Sangtin Writers and Nagar, 2006; Hesse-Biber, 2012). However, our intimate relationships with our research participants created a tension in how we present ourselves to our academic and research communities. To be clear, the tension is multi-dimensional. We have struggled with how to position our scholarly identities when engaging with our research participants, as well as how to talk about our intimate insider status when communicating to our academic communities.

In what follows, we reflect on moments of tension when we felt pulled in opposite directions after completing our fieldwork and returning to the academy. We use these moments to contribute to the ongoing conversation on feminist methodology that connects theory and practice in intimate research. In sharing our experiences as missteps and moments of clarity, this chapter considers how intimate insiders can maintain our ethical commitments to our research participants and produce effective critical scholarship. In so doing, we examine how scholars engaging in intimate research might mindfully bridge the gulf between research participant and academic perspectives. Our experiences affirm that connecting these worlds is challenging and sometimes anxiety-producing. Nevertheless, we argue that a central method of intimate research and writing should be a consideration of how to bring together these people, places and communities in order to better understand one another's perspectives and experiences.

We begin by briefly introducing our research projects and our intimate relationships with our research participants. We then transition to a series of short reflections that detail our transition from fieldwork back to the academy, and the moments of tension we felt in relation to the disparate perspectives between our research and academic communities. Although these tensions manifested differently for each of us, we highlight our strategies for navigating the distinct (and sometimes conflicting) experiences and perspectives of our research and academic communities.

Intimate insiders: Returning to the academy

Borrowing from queer theorist Taylor (2011), we identify as intimate insiders, distinguished from 'insider research' on the basis that 'the researcher is working, at the deepest level, within their own "backyard"' (2011, p.9). Taylor designates researchers who have preexisting friendships – 'close, distant, casual or otherwise' – that evolve into informant relationships as 'intimate insiders' (2011, p.8). Our research participants are enmeshed in our personal and non-academic lives, and as long-time residents of these communities, our field sites are also our homes. Our already-established membership in our field sites not only enabled our access to research participants and spaces typically closed off to outsiders, but it afforded us privileged institutional and cultural knowledge that shaped our project design, analysis and findings. We have previously reflected on our intimate insider status and our use of the feminist methodological practice of boundary-making as a means to address the complexity of intimate field sites (Cuomo and Massaro, 2016). In this chapter, we build on our prior reflections to discuss our shift from the field back to the academy, and the ethical and emotional challenges that have emerged in that transition.

Struggles associated with transitioning from the field back to academic life are common (Staeheli and Lawson, 1994; Sharp and Dowler, 2011). However, we found that our position as intimate insiders compounded these complexities. While collecting data, we consciously worked to maintain boundaries with our research participants. For us, this meant actively creating boundaries and producing emotional and physical distance in order to detach ourselves from these familiar spaces (Sharp and Dowler, 2011). Boundary-making also serves as a reminder to our research participants to view us as researchers rather than friends or family. For example, to clearly delineate moments when we were collecting data – rather than simply hanging out – we pulled out our notebooks, indicating visibly that we were taking field notes. This boundary-making practice signalled to our research participants that we were on-the-record and allowed them to consider how they might want to censor themselves for an outside audience.

Upon completing our fieldwork, we unexpectedly found ourselves abandoning the practice of boundary-making and, instead, began searching for ways to create linkages between our academic and research communities. The desire to connect the two communities emerged as we began to experience difficulties navigating their disparate perspectives and experiences. We often faced academic communities who were committed to abstract and theoretical ideals, and research participants who saw these academics as out of touch with the realities of their lives. We did not seek to reconcile the disparate perspectives, but rather we hoped to facilitate a space that supported discussion and reflection. As the common thread between the two communities, we considered how to position ourselves within both communities.

Vanessa completed a multiyear-long ethnography on the drug economy in the Grays Ferry neighbourhood of Philadelphia. Her marriage to a lifelong resident of Grays Ferry largely facilitated her access to the neighbourhood and her research subjects. While gathering data, Vanessa found it challenging to balance her fluid identity as researcher and resident. She continues to find it difficult to reconcile the perspectives of her field participants with her own academic analysis of their decisions and behaviours and its broader positioning in academic debates.

Like Vanessa, Dana also completed a multiyear-long ethnography. Her project examined the policing response to domestic violence in Centre County, Pennsylvania. Prior to beginning her doctoral programme, Dana worked for three years as a victim advocate in the domestic violence unit of a Centre County police department. She returned to the same police department to conduct research. Since completing fieldwork, Dana has struggled with how to bridge the perspectives of her research participants in the police department (who support policing responses to domestic violence) with academic critiques of state policing and mass incarceration.

In presentations of our research, we were increasingly sensitive to the divergent perspectives in the communities that we straddled. In our encounters with academics and research participants, we found it challenging to respond to critiques from either side and still remain a legitimate part of both communities. In other words, we struggled to present ourselves as credible experts and to accurately represent our research because we were so embedded in our fields. We think that this struggle is rooted in the wide gulf between the experiences and perspectives of these varied communities. As intimate insiders, we navigate a tenuous balance between two worlds that requires us to think differently about the ripple impacts of our findings, our writing and our presentations. Through discussing the challenges that we faced, we aim to generate and contribute to the conversation in feminist methodology regarding the ever complex and blurry boundary between the field and the academy.

Connecting our research and academic communities in concrete, methodical ways works to convey to the other what each community prioritizes as significant. As we change the terms of knowledge production and work to flatten the relationships of knowledge and expertise, conscious discussion of an intimate insider's role in this process becomes necessary. We offer two sets of reflections: the first, on academic presentations, and the second, on public presentations. Through these reflections, we show how we might promote greater understanding among the varied perspectives of our academic and research communities. In so doing, we seek to facilitate an intellectual space more willing to engage in candid discussions of intimate fieldwork, research participants and scholarly audiences.

Academic audiences

Vanessa begins with a reflection that explains how she mixed her personal and professional life in the field to legitimate her accounts of the drug economy in Grays Ferry. Dana then details her experiences presenting the accounts of domestic violence victims who support the arrest and incarceration of their batterers, to scholars and activists opposed to state policing and incarceration.

Bridging through disclosure: Vanessa

I gave the first academic presentation of my work shortly after completing the preliminary phase of my dissertation research on networks of care amongst men connected to the informal drug economy in Grays Ferry, Philadelphia. I presented this work initially with a great deal of apprehension. I was concerned that my discussions of Black men's efforts to create community, build stability for themselves and their families, and make complex decisions about participation in the drug economy may be read as romanticism. Even worse, I feared my efforts to contextualize violence in relation to networks of care might prove offensive to some. My findings upended many taken-for-granted assumptions about the antagonistic role of drug dealing in urban communities, and I did not know how this fact would be received by academics.

After delivering my presentation, I braced myself as hands shot up in the audience. The first question I took, from an audience member familiar with the city, challenged the truth of my narrative. He did not deem it possible that men from this neighbourhood would share such candid accounts with me. In all of my worrying, I had not anticipated this question. He could not imagine how I, as a middle-class, white woman not originally from Philadelphia (let alone the inner-city), had gained access to a neighbourhood so notoriously closed off to the outside. I felt compelled to explain that my husband had grown up in the neighbourhood and that my marriage had played a tremendous role in my ability to recruit participants and complete the ethnography. This line of questioning, which became common after my presentations, reflects the gulf between these two spaces. Academic audiences found it difficult to reconcile my life as a scholar with my life in the neighbourhood. Making clear my embedded position as a resident and family member helped to bridge the two communities, yet it required that I reveal intimate details about my life.

Although taken aback by questions from academic audiences who challenged the legitimacy of my research, I could sympathize with their misunderstandings of the Grays Ferry community, particularly with respect to residents' experiences of violence. The longer I lived in the neighbourhood, the more my understanding of violence shifted and changed. The more time I spent in Grays Ferry, the more I found myself accepting violence that, at face value, misaligned with my sense of justice. Similarly, I began to understand and even defend people's decisions to attract police harassment or maintain a public presence when it made them vulnerable to violence. These were behaviours that seemed inexplicably reckless to me at first. I wanted to convince my academic peers of my new perspective based on my experiences of the neighbourhood. Yet, as I wrote about and discussed the illegal, violent and conflicted inner lives of my research participants, on the one hand, I grappled with how to represent and contextualize how and why they made certain decisions. On the other hand, I worried about maintaining theoretical and ethnographic rigour as an academic researcher, while upholding my own ethical stance against violence. I found, as I became more embedded in life there, that the struggle to reconcile my idealized theories of pacifism with lived practice intensified.

As I sought to connect the academy with the seemingly distant lived experiences of people in Grays Ferry, I found it useful to write from the embodied stance that had been thrust upon me by academic audiences. I discovered that I had to reveal my personal relationships in Grays Ferry to discuss them honestly and represent my research clearly to my academic community. At the same time, I experienced a great deal of angst about revealing my private life in a professional setting. Nevertheless, by revealing my personal relationships, I could place myself in Grays Ferry and better demonstrate the experiences of daily life in the neighbourhood. The disclosure of my intimate relationship helped to legitimate my research findings, and it was also consistent with my feminist research methodology. To this end, I have found that being more candid about these complex relationships is necessary for intimate research and writing – however uncomfortable it may make me to reveal my private life in professional forums.

Bridging through reflection: Dana

I usually prefer small conferences. Even when the format follows the traditional and limiting fifteen-minute presentation, space tends to exist to develop relationships and continued dialogue during coffee breaks or over dinner. Consequently, I submitted an abstract for a conference on carceral geographies, organized by graduate students. The conference organizers invited a diverse group of participants, including practitioners who volunteered and worked in prisons. I felt drawn to the practitioners. These were people with backgrounds similar to mine who shared a theoretical interest in carceral geographies, but whose employment likely brought different perspectives to the conversation.

I presented midday, following a morning succession of presentations that aptly critiqued the prison industrial complex, police surveillance of youth and the disproportionate impact of policing and incarceration on poor communities of colour. I watched as like-minded academics nodded their heads in agreement to analyses rooted in Marxist, feminist, postcolonial and anti-racist theories. The small group of strong voices that led the discussions following each presentation quickly established the conference tone – reforming the criminal justice system was not an option; rather, abolitionism and the undoing of the carceral state were the only acceptable solutions to address the violence and racism embedded within the policing and prison systems. This position showed little regard for the practitioners in the audience as it spilled over into a critique not only of the power-based structures that maintained the carceral state, but also of the practitioners who worked within the institutions. I began to feel anxious and uncomfortable for the practitioners in the room. And I felt anxious about how the academic audience would receive my intimate insider perspective, which was sympathetic to the incarceration of domestic violence offenders. After only a few hours, the gulf between the academics and practitioners ran so deep that a path to productive discussion seemed beyond reach.

When my allotted time arrived, I walked to the podium and read from my prepared script. I recounted the stories of two women who unequivocally wanted the state to incarcerate their abusers; they both identified incarceration as the means to ensure their safety. I talked about law enforcement's legal obligation to arrest in cases of domestic violence, and I talked about the importance of the criminal justice system taking seriously the safety concerns of victims. I acknowledged that the two women and their appeal for punitive punishment did not universally represent the desires of all victims. I also noted that other examples in my work highlight how the state response to domestic violence can paradoxically increase feelings of fear and insecurity. I acknowledged my former employment experiences as a victim advocate as way of bolstering my claims, but out of fear of being lumped into the practitioner group and subsequently silenced, I stopped short of detailing how my intimate working relationships with law enforcement shaped my support of police involvement in cases of domestic violence.

At this conference, the tension between academics and practitioners derived from an assumption among the academics that the practitioners were too embedded to reflect on how they maintain a racist, violent institution by working in the prison system. Yet, absent from the academics' critiques of state policing and incarceration were recommendations for meaningful alternatives that could immediately address a wide range of problematic behaviours. Until my presentation, the criminalization of drugs, undocumented immigrants, and youth misconduct dominated the discussion; behaviours that we might consider victimless crimes. The problem of domestic violence offered a moment for pause and reflection that underscored the gulf between abstract ideal and everyday practice as I detailed the urgency of how a police response to a 911 call for help might mean life or death for the victim.

As a scholar who has spent ten years either working directly alongside or studying the police, I still hesitate when I am within academic circles to express my support of police involvement in domestic violence cases. I am especially reluctant to fully position myself in relation to my friendships with my former police officer colleagues, or express the sympathy that I feel for police officers who respond to domestic violence incidents. This is not because I believe that my intimate insider status discredits my research, but because of the limited opportunities to write and present in ways that allow for the space or time to explain positionality. For this conference, to ease the tension and bridge the gulf between the academics and practitioners, I emphasized my expertise as both a former victim advocate and domestic violence scholar. I catered to both audiences and acknowledged the limitations inherent in the policing response to domestic violence, but also stressed the need for viable alternatives. The tension in the room did not fully dissipate, but my intimate insider perspectives created a small space for further reflection on the necessities of bridging theory and practice.

Blended audiences

In our next set of reflections, we discuss our experiences presenting and working with combined audiences of academics and our research communities. Vanessa begins by detailing her experience returning home to Grays Ferry to complete research with students and her former research participants who transitioned into paid research assistants. Dana then describes a public presentation that she gave on her research with police officers and academic colleagues in attendance.

Bridging through encounter: Vanessa

In the summer of 2015, I returned to Grays Ferry to continue my research on the informal economy. More specifically, I wanted to examine the impact of incarceration on household economies. This time, I returned with three students to complete household surveys. Bringing outsiders to engage with the community added another dimension to my already complicated role in Grays Ferry. Previously, my research position had been relatively contained: I was mainly an observer, conducting interviews with select individuals over the course of two years.

For this project, we used my home in the neighbourhood as a base. I paired my students with residents of Grays Ferry who were past participants in my dissertation research. While a great deal could be said about this balancing act, here I focus on how creating a domestic space that welcomed these varied people, their experiences and their anxieties was a primary, albeit problematic strategy to bridge these communities. None of my students were from Philadelphia, and they had little experience with incarceration and even less experience with field research. The residents who worked with us had little experience with college students, teaching or field research. My home became the shared space in which to bring these communities together.

When I interviewed people in Grays Ferry years earlier, I worked to stage an interview space that would help create a boundary between us. In this new situation, I arranged my home to connect my students to my neighbourhood researchers. I (mostly unconsciously) worked to strike a balance between a professional and a domestic space that would comfort my visitors as we built a small community of researchers. In this space, we were changing what it meant to be a researcher as we combined the knowledge and skill sets of college students with the knowledge and skill sets of Grays Ferry residents.

Our first day of survey training was filled with apprehension as the students met the research assistants from Grays Ferry. As neighbours and family friends, the research assistants felt at ease in my home, at least at first. They entered this space comfortably as a reflection of their familiarity with my husband, eight-month old son and the day-to-day rhythms of neighbourhood life. However, when they saw the chairs and the PowerPoint set-up, the researchers from Grays Ferry teasingly expressed their apprehension. One joked he would just play with my son outside instead. Another asked how long this would take, and what they would have to do. Socially they were at ease in my home in a way my students were not. The students, instead, were comfortable with the classroom-training environment, the computers for data analysis and the concept of research. I faced the challenge of helping the students adapt to the neighbourhood rhythms and helping the residents adapt to the research rhythms.

Over the course of two weeks, the hybrid space welcomed my neighbours and students, and their varied experiences. We shared meals, reflected on our days and enjoyed the company of one another, connecting over a shared sense of care about this neighbourhood. In time, my neighbourhood researchers became more comfortable with survey design and had discussions about sample validity, response rates and ways to improve future surveys. My student researchers became more enthusiastic, immersing themselves in the neighbourhood as they realized, with excitement, how tightly-knit a neighbourhood it was, and as they jokingly learnt slang from their neighbourhood researchers.

Conducting the survey was a culmination of several years of work in Grays Ferry, and I felt proud of what we had accomplished as a team. Those few weeks, however, remain etched in my mind for another reason: I had reached a level of exhaustion and vulnerability that I never experienced professionally. As my students and Grays Ferry residents were partnered together, they spent time beyond my purview. As a result, I could no longer control how my students and neighbourhood researchers perceived the other or how they perceived me. Surely this is always the case, but my concern in this context derived from the fact that I was no longer physically present to mediate their interactions as they completed their work independently. I constantly worried that one may offend the other in a way that could potentially damage the carefully crafted relationships I had formed with both groups. My angst revealed clearly to me how scholars must constantly perform for varied audiences and how complicated it can be to disrupt the delicate balance by mixing them together.

As my position shifted and I took on new roles within the academy, I found it necessary to reflect on what it means to be an intimate insider who represents the bridge between Grays Ferry and my university. This survey project was laborious, complex and risky for me personally, and it required reflexivity and discussions with both my scholarly and field-based communities. It is clear this connection – between practice and theory, between field and academy – is important for creating an accessible and relevant academy. In the end, however vulnerable the process made me feel, I also felt the process of my field relationship was made more transparent to my students. More importantly, for my research partners in Grays Ferry, it was a significant entry for them to develop a more nuanced and transparent understanding of my role as a scholar. Bringing students to Grays Ferry built and solidified their connection to research and knowledge production. The connection between the students and the residents of Grays Ferry remains. They stay in contact and ask about the other when they see me. Creating these small relationships and connections is a fruitful step in opening the academy. Further, both groups inquire about the data, the findings and what I am doing with their work in a way that demonstrates this partnership created a longer term vested interest for both groups.

Bridging through empathy: Dana

The State College branch of the Association of American University Women (AAUW) sponsored a community presentation I gave following the completion of my fieldwork. With two domestic violence homicides within my field site in the previous year (one of which occurred after the batterer's arrest), my research on the county's policing and prosecution response to domestic violence garnered significant interest.

AAUW volunteers set up rows of seats in a reading room at the public library. The library is centrally located in town, a short walk from campus and next door to the police department where I conducted research. The rows quickly filled as I chatted with my former victim advocate colleagues and watched an assistant district attorney, patrol officers, detectives, and family law attorneys file into the room. They took seats among a gaggle of students from a women's studies class, faculty from various departments, survivors of domestic violence and dozens of members of the general public. I had anticipated a blended audience, but the complexity of executing this talk materialized when I saw the chief of police sitting next to a women's studies professor!

I took a deep breath and introduced myself to the audience of familiar faces. I acknowledged the research participants in the room and thanked them for attending, affirming that their presence attested to their commitment to actively address the problem of domestic violence. I then joked that their attendance might also relate to feelings of anxious curiosity for how I would represent them in my writing and presentations. The nervous laughter that followed made clear that my joke hit a nerve, and I suddenly understood the amount of power that I held as a researcher over my participants' emotions.

As practitioners, we routinely evaluate our community response to domestic violence; yet those internal reviews are seldom circulated. My presentation provided a rare public critique of a law enforcement process close to the hearts of those in attendance. With the recent homicides having occurred on law enforcement's watch, I realized they were anxious about how I might evaluate their practices. The blended audience heightened their stress, and I felt empathy for them. For the remainder of the presentation, I attempted to offer a balanced critique that avoided causing angst for my research participants, while providing critical feedback to inform practice, policy and the general public. As a scholar, I felt an obligation to represent my findings accurately to the audience. As a former victim advocate, I felt protective of my friends and colleagues. These are not necessarily competing interests; however, it became clear to me during this presentation that my intimate insider identity caused anxiety for my research participants, who were hearing, for the first time (and in a public forum), my academic analysis of their perspectives.

The venue offered a unique opportunity to bring different types of people together who infrequently cross paths. As an intimate insider, this presentation became a way to fulfil an ethical commitment to my research participants by making visible how I represented their perspectives to academic audiences. For my academic colleagues and others critical of policing, the shared space offered an opportunity to humanize law enforcement. After the presentation, the questions I was asked made evident the collective concern for addressing the problem of domestic violence and it became clear to everyone in the room that the overlapping audiences shared more in common than we realized. When the event concluded, the police chief approached me and kidded, 'You could have been tougher on us. Maybe next time.' I appreciated the chief's commitment to public accountability, and knew that the comment also signalled that my position as an intimate insider remained unchanged. I explained in response that I did not intend for the presentation to be a lopsided critique of law enforcement. Rather, the presentation represented an opportunity to bridge two audiences, allowing each an opportunity to reflect on their critiques of each other.

Conclusion

Upon completing our fieldwork, we faced the common challenge of reconciling our insights from the field with the demands of academic rigour. Our positions as intimate insiders exacerbated this challenge in the ways that we discuss above. However, we found that intimate field research offers a unique perspective from which to consider bridging these two distinct worlds. As we have navigated the shift from fieldwork back to the academy, we found research and presentation served as a way for us to bring our academic and research communities into literal and metaphoric conversations with one another.

Bridging serves to further the accessibility of our scholarship and make our standings transparent to each community, fostering more ethical, inclusive research interactions. When our research participants are able to experience what we do as scholars, they are positioned to navigate their participation in our research more effectively. They are more informed by knowing what we say to our academic peers and, within that process, we are challenged to be more mindful. Although the position of intimate insider is difficult and fraught at times, we would not conduct research differently. We remain committed to doing and supporting embedded, intimate research, and we encourage the continued search for ways to bridge these perspectives with our academic colleagues, even if it means that we navigate more tension than we expected.

Works cited

Cuomo, D. and Massaro, V.A., 2016. Boundary-making in feminist research: new methodologies for 'intimate insiders'. *Gender, Place and Culture*, 23(1), pp.94–106.

Hesse-Biber, Sharlene Nagy, 2012. *Handbook of feminist research: theory and praxis*. Thousand Oaks, CA: SAGE.

Sangtin Writers Collective and Nagar, Richa. 2006. *Playing with fire: feminist thought and activism through seven lives in India*. Minneapolis: University of Minnesota Press.

Sharp, Joanne and Dowler, Lorraine, 2011. Framing the field. In: Vincent J. Del Casino, Jr., Mary Thomas, Paul Cloke and Ruth Panelli, eds., *A companion to social geography*. Oxford: Wiley-Blackwell. pp.146–60.

Staeheli, Lynn A. and Lawson, Victoria A., 1994. A discussion of 'women in the field': the politics of feminist fieldwork. *The Professional Geographer*, 46(1), pp.96–102.

Taylor, Jodie, 2011. The intimate insider: negotiating the ethics of friendship when doing insider research. *Qualitative Research*, 11(1), pp.3–22.

Part II

Emergent effects of including one's own story

6 Intimate creativity

Using creative practice to express intimate worlds

Clare Madge

Introduction

In contributing to the intimate turn in geography (Valentine, 2008; Price, 2013; Pain and Staeheli, 2014; Smith, 2016), this chapter explores how creative practice might be used to contemplate geographical questions of intimacy. In particular, I consider whether creative practice might be a means of 'writing ourselves differently' (Tamas, 2009, p.1) in this move towards intimacy. Below I deliberate several questions. What are some of the potentials and limitations of creative practice for expressing intimate worlds? What challenges does intimate creativity pose for analytical strategies? And how might such analytical quandaries be traversed?

In contrast to intimate writing based on prose, I examine these questions through the frame of creative bricolage, a do-it-yourself form of creative practice which improvises, reuses and recombines available materials. In the chapter I employ two creative bricolages (Figures 6.1 and 6.2), which both consist of intimate poetry and photography. I use these figures as a prism to reflect on whether the creative practice of bricolage offers potential for geographical studies of intimacy. Initial thought suggests that poetry lends itself particularly well to intimate expression, for through its compressed form it can convey complex emotions and worlds in a few words. Poetry is also a means to express everyday intimacies, as well as extraordinary ones, through imaginative projections. However, in an active intention to disrupt the scriptural/visual divide, the creative bricolages also consist of a manipulated photographical image superimposed over the poem, to see what generative encounters, what novel ways of thinking might result from this hybrid creative form.

In employing creative bricolages of poetry and photography, I aim to make a contribution to the burgeoning literature on the creative (re)turn in geography (Hawkins, 2013; Marston and de Leeuw, 2013; Hawkins and Straughan, 2015). In particular, I am responding to Crang's (2014) provocation for geographers to think carefully about this creative (re)turn, specifically considering how it can open up intimacies and perform previously hidden worlds. To this end, rather than looking at creativity as something outside of myself, I write this chapter as a creative agent, as both the subject and maker of the creative bricolages. This is an approach which transcends dualisms of author/object, insider/outsider and public/private.

The creative bricolages are based on two specific (and some might argue extraordinary) moments in the life of my 'minded-body' (Hayes-Conroy, 2010). These are experiences that occurred with an intervening gap of over a decade. Figure 6.1 revolves around my pregnant minded-body and it comes from an unpublished paper originally written in 1998 that explores aspects of my pregnant minded-body performing geography. In the image I am trying to convey my wonder at my growing minded-body and how this led me to consider that 'space is not a fixed entity'. The set of associated poems are passionate expressions of being pregnant but they also communicate the profound questions pregnancy imposed on my intellectual understanding of boundaries, borders and interrelationships (in this case specifically between mother and child).

Figure 6.2 is a reflection on my ill minded-body and it originates from a book of cathartic creations that I made whilst being treated for breast cancer in 2009. I used these cathartic creations in the process of attempting to fathom the intricacies of facing a life-threatening illness, as I was passing through it, with an unknown end-outcome. These creations facilitated a vital process of catharsis that enabled me to express the unspeakable and engage in reverie (Tamas, 2014, p.91), helping me to process and contemplate my cancer diagnosis and treatment. The focus in figure 6.2 is on chemotherapy and the photographic image is a spectral, inverted collage of the numerous medications I was taking during chemotherapy treatment. I wrote the poem immediately after a session of chemotherapy, while the medicine was still 'pulsating fluorescent through my cells' and my mind was wandering to 'dreamy, creamy, hazy places' owing to steroid medication. The poem helped me to externalize the visceral pain I was feeling and move towards a place of (emotional, physical, spiritual) recovery. The creative bricolages in figures 6.1 and 6.2 are therefore deliberately intimate and they paint a picture from the inside, from one minded-body going through these particular experiences of pregnancy and chemotherapy treatment.

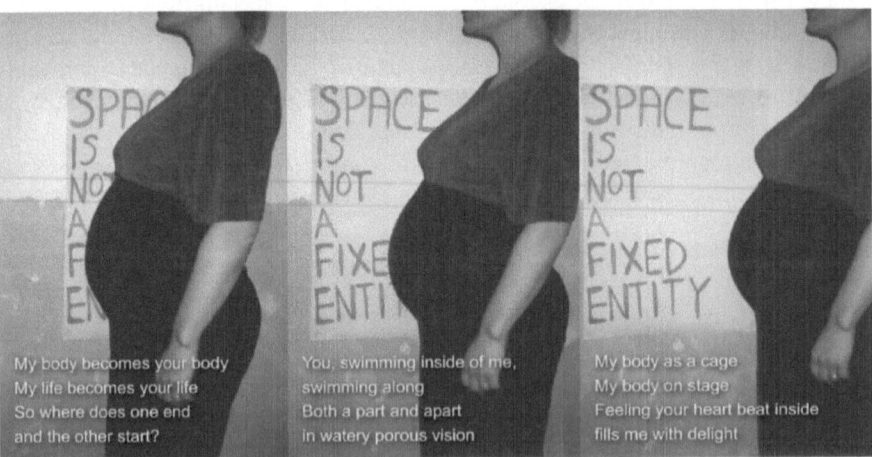

Figure 6.1 Pregnant possibilities: Space is not a fixed entity

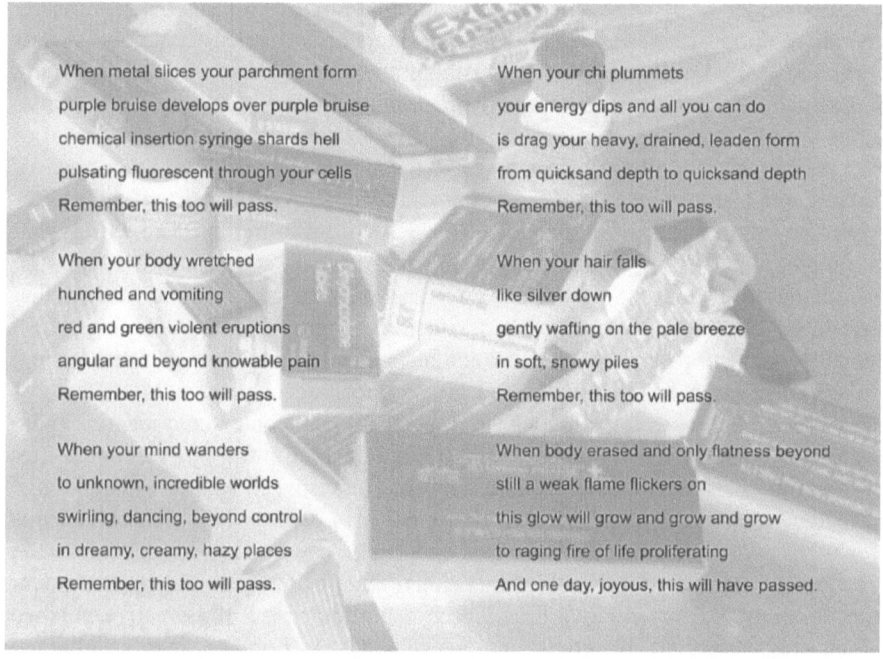

When metal slices your parchment form
purple bruise develops over purple bruise
chemical insertion syringe shards hell
pulsating fluorescent through your cells
Remember, this too will pass.

When your body wretched
hunched and vomiting
red and green violent eruptions
angular and beyond knowable pain
Remember, this too will pass.

When your mind wanders
to unknown, incredible worlds
swirling, dancing, beyond control
in dreamy, creamy, hazy places
Remember, this too will pass.

When your chi plummets
your energy dips and all you can do
is drag your heavy, drained, leaden form
from quicksand depth to quicksand depth
Remember, this too will pass.

When your hair falls
like silver down
gently wafting on the pale breeze
in soft, snowy piles
Remember, this too will pass.

When body erased and only flatness beyond
still a weak flame flickers on
this glow will grow and grow and grow
to raging fire of life proliferating
And one day, joyous, this will have passed.

Figure 6.2 Chemo mantra: this too will pass

The remainder of the chapter is divided into four parts. First, I contemplate what insights might be gained from using creative practice to express intimate worlds. Second, I consider some limitations to such intimate creativity. Third, I reflect on two strategies to analyse intimate creativity: corporeal extrapolation (scaling up from the intimacy) and imaginative transgression (utilizing creative practice to generate new geographical insights). Fourth, I finish the chapter with some conclusions, arguing that richly evocative creative expression can fruitfully add a further dimension to the diverse range of intimate writing practices used by geographers.

Making the case for intimate creativity

Recently there has been a proliferation of interest in alternative creative writing strategies in geography, including experiments in poetry, theatre plays and performative/collaborative writing (de Leeuw, 2012; Cook, 2014; Richardson, 2015). In this chapter I focus on creative practice (specifically bricolages of poetry and photography in combination), to consider five key contributions that this alternative form of expression might provide for studies of intimacy.

First, creative practice can be used as a form of embodied storytelling, being a means to express corporeal intimacies. In this instance, the inclusion of the pregnant and ill minded-body can potentially shed some light onto the intimacies of these specific minded-body experiences. For example, in figures 6.1 and 6.2, I am using my minded-body as a research tool which counters disembodied accounts of maternal and mortal events and gives some insight into the everyday experiences of being pregnant and undergoing chemotherapy treatment. Intimate creativity can therefore act as a portal into inner worlds, bringing alive a fleshy, fine-grained portrayal of everyday lived life, albeit from a specific framing or perspective. Such creative practice can also produce insider accounts of specific experiences that might otherwise be too difficult or too unethical to research, revealing intimate worlds that might otherwise remain hidden. Employing a creative practice that speaks through the minded-body consequently opens up a discursive space for minded-bodies that have hitherto been muted or forcibly silenced in geographic texts, enabling different stories to be told about the world that may not be accessed through other writing strategies. In figure 6.2, for instance, I give a poetic account of chemotherapy, as seen through the 'eyes' of someone (in a specific social context and location) who is currently receiving treatment, offering unique insight into living through a life-threatening illness as it is actually experienced. The visual encounter with the range of medications in the photograph in figure 6.2 also makes vivid elements of this storying experience. Together the text and image can be layered together in the creative bricolage to reveal multi-dimensional features of chemotherapy treatment as it is endured. Thus creative bricolage is a particularly useful creative practice which can recombine materials in unusual ways to bring intimate, embodied experiences to life.

A second potential of creative practice is that it can be a way to gain insight into emotional intimacies: it can be a way of encountering the intellect through the heart. Through my creative bricolages, I make the case for tender, fleshy, intimate expression which can breathe life into the humanity of the lived experience of pregnancy or a cancer diagnosis, allowing for a visceral resonance. Indeed, sharing intimacies through creative practice has the potential to initiate a process of rapport building which can become the basis of shared understanding. Figure 6.1, for example, illustrates that intimate creations can have an emotional register. The poems express intensive passions and sensibilities – 'feeling your heartbeat inside fills me with delight' – and so have potential to evoke a closer emotional relationship forged through an empathetic response. Hence, creative practice may be a means to access emotions and empathies, giving insight into the multiple (sometimes painful) experiences of life over an emotionally-sanitized version (Madge, 2016). Creative poetic expressions of the self can also locate the researcher emotionally, presenting a counterpoint to detached, disembodied, unemotional geographical accounts. Using a creative practice that 'speaks through the body' (Duffy, 2013) can therefore produce passionate accounts, which can enable emotional reverberation, allowing the geographical cannon to be expressed differently.

Third, creative practice may also hold potential to capture the transitory and relational nature of intimacy. Intimacy is rarely fixed; rather it is transient, varying in and between relationships, groups of people, various things and different practices, all of which change over time and vary with place. In the case of figure 6.1, this relational nature of emotional intimacy between mother and child is expressed in the poem at several points: 'my body becomes your body/my life becomes your life/so where does one end and the other start?' while the repeated photographic images are presented to capture a sense of the movement and change to my corporeal contours, illustrating that the pregnant minded-body 'is not a fixed entity.' The bricolage in figure 6.2 also illustrates the relationship of my ill minded-body with things (medicines, syringes, vomit) and practices (cannulation, daydreaming, hair loss). The creative bricolages in figures 6.1 and 6.2 thus usefully demonstrate the changing, relational nature of intimacy. Indeed, different aesthetic forms may also be able to access other intimacies in a manner not possible through more conventional writing strategies: for example, physical intimacy might be felt through touching sculpture while experiential intimacy may be accomplished through the creative process of making aesthetic work. In this manner creative practice can bring to life the multi-sensory character of intimacy.

Fourth, in some circumstances creative practice can enable cathartic release. Employing intimate creative formats might enable those going through specific experiences make some sort of imaginative cathartic move through the creative process. This creative agency might be a way of writing oneself (back) into being (Philo, 2014, p.285), of bearing 'witness to occurrences that cannot be understood or experienced in any other manner' (Tamas, 2014, p.91). This was certainly the case in producing both of my creative bricolages. Figure 6.1, for example, enabled me to work through the profound changes occurring to my bodily contours when pregnant and my consequent intellectual conceptualization of space (more on this later), while figure 6.2 was a means to externalize the manifold emotions and visceral pain surrounding chemotherapy treatment. Enabling catharsis in the process of creative-making, as a component of the research process, however, requires mature research expertise, well-developed emotional intelligence and probably psychotherapeutic or art therapy training and extremely careful handing, but it might ultimately enable living with and transforming trauma (see Madge, 2016).

Finally, creative practice can also be used to explore imaginative intimacies, through its transformative potential to enact the world in novel and surprising ways. As creative expression can be startling or evocative, it can open up enquiries into intimacy in exciting and unanticipated directions. It can encourage the exploration, articulation and investigation of more strange, unforeseen and unique worlds which ignite the imagination. Here bricolage as a specific form of creative practice can be particularly fruitful since it allows for an improvisation and experimentation. In the case of figure 6.2, I had no pre-determined plan of what I was going to produce for this chapter, the creative moment just spontaneously took me over as I montaged together the photograph and poem. I find this form of experimental creative practice most liberating in contrast to more predictable academic writing conventions. This commitment to unknowing that creativity fosters is best summarized by this wonderful quote by Tamas (2009, p.20):

I might use creative methods, not in order to be clever, but because I myself don't know the story that is sliding around in me, looking for an opening. ... At its best, [art] can show – not tell – us something about what it is to be human.

Having outlined a few of the potentials of creative practice for producing embodied, emotional, relational, cathartic and unexpected accounts of intimacy, I now go on to examine some of the limitations associated with intimate creativity.

What limits to intimate creativity?

One issue of concern surrounding intimate creativity, which particularly arises when using creative practice to express the self, is that care must be taken that warm and friendly encounters do not drown out other more troublesome forms of intimate expression, such as the silent, the inexpressible, the angry or the colloquial. Oppression is, after all, very intimate (Verhage, 2014). It is important that the focus on the intimate does not become a parochial snapshot of a very specific subset of our diverse world, which misses the bigger economic and political picture and reveals only certain people, places and experiences. The question thus arises: to what extent is the intimate creativity expressed in figures 6.1 and 6.2 useful in responding to multiple voices or how and why might it just become a tool for privileging my particular world view? Intimate creative practice is potentially just as susceptible to asymmetrical power relations, dominant voices and exclusions as any other mode of knowledge creation, requiring constant vigilance and attention to the 'power that swirls around us' (Moss, 2014, p.803). It is important to foster intimate creative practice from variegated worlds, validating stories of multiple intersecting regional, national, linguistic, ethnic, racial, class, age, sexuality, gender and political asymmetries, while always being vigilant to issues of disclosure, exposure and epistemic violence in the process (Sharp, 2014). I would therefore take a cautious stance. While being conscious of the struggle feminists have had to counter the charge that intimate writing is narcissistic and self-indulgent (Moss, 2014), which can marginalize intimacy to a 'supporting role' (Pain and Staeheli, 2014, p.345), there is also an imperative to avoid uncritically valorizing intimate creativity as a tool of embodied storytelling that gives insider accounts, by asking whose bodies, which insiders, what intimate stories and for what purpose.

A second concern revolves around the limits and risks of creative practice to express emotions and evoke empathy. While intimate creativity can have a productive emotional register, one of its difficulties revolves around the limits to empathy. Take figure 6.1, for example. The inclusion of my pregnant minded-body in the image raises several troublesome issues. Using the image of my pregnant minded-body is an intensely political act, placing the female firmly within the genealogy of geography and insisting that pregnant bodies are an important element of geography's story. Pregnancy is, after all, a very common human experience occurring throughout the world (Noxolo, Raghuram and Madge, 2008, p.150).

However, pregnancy is also an occurrence that is often accompanied by loss, pain, violence or unachieved potential (Madge, Noxolo and Raghuram, 2004). It also has the potential to exclude, through the inability to become pregnant, through unwanted pregnancy, through active choice to not become pregnant or through normalizing certain heteronormative expectations/experiences. Hence empathy surrounding pregnancy can never be a guaranteed outcome of this particular creative expression of intimacy. Indeed, intimacy implies an inter-personal, reciprocal relationship but it is important to reflect on the extent to which it is possible to evoke empathy and dialogue in the intimate encounter if there is no shared experience or understanding.

These contestations over creative practice can also be considered at the boundaries of the intimate exchange. Although creative practice has potential to enable cathartic release, what happens when intimacy starts fraying at the edges, when the intimacy becomes too intimate, too unbearable, too excessive, inappropriate or embarrassing? What about those intimacies that are forgotten, misremembered, or only shared selectively (Sharp, 2014) or those that cannot be acknowledged in private, let alone spoken about in public – those intimate non-truths, silences and refusals? After all, intimacy is not always positive and generative; it can be destructive, uncomfortable and dysfunctional too, involving violence from others (Pain and Staeheli, 2014) and processes of distancing (Price, 2013). Indeed, emotional boundary-making and detachment may be an important methodological strategy for 'intimate insiders' undertaking research in 'familiar fields' (Cuomo and Massaro, 2016) or by which oppressed collectivities might regroup and 'hear oneself talk' (hooks, 1989, pp.5 and 31; see also Verhage, 2014).

Certainly some topics probably lend themselves more fruitfully to an intimate creative encounter than others. For instance, is the scriptural poetic encounter with my pregnant minded-body in figure 6.1 less disconcerting than the cathartic engagement with my bruised, drug-induced, minded-body during chemotherapy in figure 6.2? Or is the visual encounter with abstract packets of pills and ghostly bottles of medicines in figure 6.2 less troubling that the direct engagement with my actually pregnant body (as was a decade ago) in the embodied photography of figure 6.1? I cannot predict the answerer to these questions as I do not know what response the figures might (or might not) evoke, so intimate creative practice, especially about difficult or traumatic events, is risky, with unforeseen outcomes. There is much work to be done to examine more fully how makers of creative works might negotiate the inevitable risks, ethics and exposures of intimate research and the reactions, significance and repercussions on their audiences. For example, in figure 6.2 there are many risks involved in attempting to convey the surreal and intangible nature of chemotherapy, and the sheer grit required to endure it. Some people reading this chapter may have experience of cancer (either personally or having a family or friend who is living with cancer, living through it, or has sadly died from it), or other life-threatening illnesses and losses, so figure 6.2 may be an emotionally demanding for some, triggering memories of loss and feelings of grief (see Madge, 2016). I am therefore troubled by publishing this image and the (varied) emotional responses it might trigger in the audience.

Other risks revolve around myself and include questions such as: what will people make of my creative bricolages? Will expressing myself through creative intimacy open me up to more criticism than a more conventionally written chapter? What risks and vulnerabilities (emotional, legal, in terms of academic credibility) might be at stake in using my experiences as the focus of the creative bricolages? How might I, and others close to me, feel about these revealed intimacies in the future? And how, and why, might the focus on my experience of breast cancer and pregnancy reduce, misrepresent or obscure the experience of others, running the risk of reinstating an undifferentiated universal minded-body? There are no easy answers to these tricky questions, for as Hawkins (2011, p.472) notes, 'it can be hard to study, and even harder to write about, these personal and experiential ways of knowing'. That said, perhaps it is most useful to consider how such challenging questions are played out, and worked through, in specific contexts/locations through the process of particular research projects?

Having considered some of the more unsettling aspects of intimate creativity, I now move on to contemplate two strategies that might be used to analyse intimate creativity: corporeal extrapolation and imaginative transgression. This focus on analytical strategies is important, for while much attention has been paid to different writing strategies used to express intimacy, there is less reflection on how to analyse these alternative approaches. Below I reflect on two strategies through which creative practice might be mobilized as data to provide insight into the intimate worlds of pregnancy and breast cancer.

Analysing intimate creativity: Some reflections

Corporeal extrapolation

Although my intimate creative bricolages present personal, individualized accounts of pregnancy and chemotherapy treatment, how possible is it to scale up or extrapolate out from these unique corporeal experiences? How can intimate creativity move beyond personal catharsis to hold wider analytical resonance, value and worth, and in asking this question, what god-trick am I falling into? Although intimacy often appears on first sight micro-spatial/physically proximate, many have demonstrated that intimacy consists of entanglements of proximate and distant intersecting relations, interactions, emotions and practices (Kraftl, 2013; Pain and Staeheli, 2014). Pratt and Rosner (2012, p.3), for example, argue that 'intimacy does not just reside in the private sphere; it is infused with worldliness.' Hence if scales are mutable and dynamic, as much conceptual and socially constructed as having material effects, following Kraftl (2013, pp.178 and 180), I want to suggest that intimate creative expression can be spatialized beyond immediacy.

Here I am thinking about the minded-body acting as a site (a gathering of flesh, emotions, discourses, practices, relations with others) through which the intimate can be lived and through which the embodied event of my creative bricolage takes place. Intimacy is continually reconfigured through these constellations of emotional, corporeal and material practices and relations (see Andrucki and Dickinson, 2015, p.5). Intimacy is thus most usefully conceptualized as circulating in (and between) bodies in a variety of ways as they move through and perform intimate acts in various times and places (see Andrucki and Dickinson, 2015, who have used a parallel argument to conceptualize a performative reading of centres and margins). This pregnant and/or cancerous minded-body that is performing intimate acts is a site that resonates and endures: these bodies are an important component of the geographical landscape that are unlikely to go away.

Therefore while figures 6.1 and 6.2 capture two specific moments in the life of my particular minded-body, pregnancy and breast cancer are also experiences which hold worldwide relevance. So while the creative intimate sources are clearly partial and situated, they also portray a picture of the emotions and visceral corporeality of certain experiences that must have broader appeal. In figure 6.2, for example, I write from the frame of someone experiencing one specific illness, with a particular knowledge of that illness, located in a precise place, with its particular system of health care provision, embedded in specific social and political networks and experienced through a distinct minded-body. Breast cancer is, however, a prevalent worldwide disease that is a part of many people's lives. As a health issue of (differential) global significance, it is a topic of important consideration. Latest World Health Organization statistics demonstrate that breast cancer is by far the most common cancer in women worldwide, with 1.68 million new cases diagnosed in 2012. This accounts for nearly one quarter of all cancers diagnosed in women; it is also the most frequent cause of cancer death for women in 'developing' regions and the second most common in 'developed' regions. In 2012 it was estimated that 522,000 women died from the disease, with mortality rates varying across the world (http://www. cancerresearchuk.org/cancer-info/cancerstats/keyfacts/breast-cancer/cancer stats-key-facts-on-breast-cancer). Given the significance of this disease, which has such devastating multifaceted impacts, clearly writing about breast cancer as an intimate insider holds value and wider significance and relevance, particularly to counter the abstract, disembodied nature of such medical statistics. Here I am making the case for the potential of intimate creativity to travel through corporeal extrapolation but this is a claim based on resonance, not universalism (see Pratt and Johnson, 2014).

Hence, in using my intimate embodied reflection on chemotherapy in figure 6.2, it is possible to extrapolate out from the creative bricolage, using it as a pivot to consider broader questions about the experience of living through treatment for breast cancer. This is partly because such intimate experiences of (im)mortality locate us in a web of relations with others (family, friends, medicinal/therapeutic practitioners, designers of medical equipment) and broader contextual factors (geopolitics of health care, cultural issues influencing treatment and diagnosis, wider discourses, knowledges and intellectual regimes surrounding cancer), in which there is 'a mutual enfolding of self and world that inevitably moves us beyond the singular personal experience' (Hawkins, 2011, p.467). Thus since intimate creative practice is rarely only about the self, but is often permeated with significant others and embedded within a wider social, political and economic matrix, it can be utilized both as explicit testimony but also as a fulcrum for broader critical reflection. In this manner, creative practice may be used as a pivot which enables corporeal extrapolation from the intimate, overcoming critiques of parochialism and self-referential catharsis. In knitting together both the small story and big picture, creative practice can enable an intermeshing of intimate expression and a wider political exteriority, challenging spatial hierarchies that bind intimacy *only* to the body and enabling accounts in which the proximate and distant become indivisible (see Pain and Staeheli, 2014).

Imaginative transgression

A second strategy to analyse intimate creativity is to think about it as an experimental practice which can act as a catalyst for innovative thinking and debate. Creative bricolage, for example, might be considered an ongoing creative process of thoughtful-making, of patching together and recombining materials to express new and imaginative ideas. Rather than using a positivist notion of reliability and validity of data, perhaps the rigour and credibility of creative practice is related to its ability to bring an experience to life, to communicate emotions, to its ability to 'do something' to the audience (cf. Madge, 2014)? For example, the figures I have used in this chapter include both poems and photographic images. These creative forms have been harnessed as a route into imagining the corporeal worlds of pregnancy and breast cancer. Both figures jar against accepted ways of articulating geography in the attempt to spark the imagination through evoking the unexpected and creating fertile imaginings of important women's worlds that often remain under the geographical radar. Such intimate creative expression can be used to make metaphorical and imaginative leaps or transgressions, through which dominant meaning might be transformed.

For instance, the poems in figure 6.1 delight in the profound challenge being pregnant had for the way in which I theorized space. The experience of being pregnant illustrated the complexities of space: space is not a fixed entity, as my changing corporeal contours testified; space is rarely immutable because boundaries between the mother and child are porous, suggesting margins can be disrupted and transgressed. Paradoxically, space can also be conceived of as both a part of and apart from (i.e. a baby inside your pregnant minded-body is both a part of you – literally – and apart from you as a separate human being), suggesting the synchronicity of co-existing spaces occurring at various interior and exterior scales. Expressing intimate worlds through creative practice can thus be a means to unsettle and reassemble conceptualizations of space (or other geographical concepts). In this example, figure 6.1 suggests a relational mapping of space whereby different spaces can exist beside one another, which helps create a way of thinking about space that is less hierarchical and allows for multiplicities of co-existing spaces. The pregnant minded-body can therefore become a metaphor for a reconstructed way of thinking about space, and it is through intimate creativity that such imaginative transgression can occur that pushes at the boundaries of geographical knowledge. Hence creative practice might fruitfully be utilized as a telling case which 'serves to make previously obscure theoretical relationships suddenly apparent' (Sheridan, et al., 2000, p.14).

This insight suggests that by sparking the imagination, by speaking through the minded-body, and a particular type of minded-body at that, intimate creative practice can address that which is still often deliberately left out (or conveniently forgotten) in particular geographic texts. The use of the photograph of my pregnant minded-body in figure 6.1 is an attempt to raise questions about the geographical gaze; it is about challenging the fixed binary of outsider and insider as my minded-body becomes subject and object of the photo and it is making a case for bodied terrains to enliven our understanding of geographical space (Madge, Noxolo and Raghuram, 2004). Through a celebration of the landscape of intimacy, the image strives to examine the possibility of a feminist politics of visual pleasure (Nash, 1996, p.149), in which the 'already-thereness of the intimate' is asserted (Pain and Staeheli, 2014, pp.345 and 346). Thus in challenging geography's discursive boundaries through imaginative transgression, I make a claim for the transformative potential of creative practice to shift the existing terrain of geographical knowledge and provide new exciting visions.

Conclusions

This chapter has considered the use of creative practice to contemplate geographical questions of intimacy. The focus has been on creative bricolage. Through corporeal extrapolation and imaginative transgression, the chapter has illustrated that richly evocative creative expression can fruitfully add a further dimension to the diverse range of intimate writing practices used by geographers. Creative practice has potential to produce emotional and embodied accounts of intimacy, while also revealing its dynamic, relational and multi-sensory nature. Creative practice can also be useful in allowing for catharsis and the production of unexpected and unusual accounts of intimacy. However, while I am attracted to intimate creativity, I am troubled by it too. It is clear that a creative disposition does not simply offer an innocent, benign solution to masculinist writing practices. It would be naïve to uncritically valorize intimate creative practice and believe that it can automatically overcome the complex layerings of power and politics involved in geographical knowledge construction. The employment of intimate creative practice does not automatically become either a route into a more accurate portrayal of geographical worlds, emotions and intimacies, nor does it simply enable more cathartic and visionary expressions: there are limitations and risks involved in any creative practice. While being mindful of such limitations, the chapter has shown how bricolage in particular might be an effective creative practice for revealing geographic intimacies, having the ability to recombine materials in unusual ways to bring the multi-dimensional nature of intimate experiences to life.

Works cited

Andrucki, Max and Dickinson, Jen, 2015. Rethinking centers and margins in Geography: bodies, life course, and the performance of transnational space. *Annals of the Association of American Geographers*, 105(1), pp.203–18.

Cook, Ian, et al., 2014. 'Afters': 26 authors and a workshop imagination geared to writing. *Cultural Geographies*, 21, pp.135–40.

Crang, Mike, 2014. Societies/economies/cultures panel. [presentation]. 6th Doreen Massey Annual Event: *Provocations of the present: what culture for what geography?* Open University, London, UK, 6 June. Available at: http://stadium.open.ac.uk/stadia/preview. php?whichevent=2425&s=1&schedule=3133&option=&record=0. [Accessed 15 August 2016].

Cuomo, Dana and Massaro, Vanessa, 2016. Boundary-making in feminist research: new methodologies for 'intimate insiders.' *Gender, Place and Culture*, 23, pp.94–106.

de Leeuw, Sarah, 2012. *Geographies of a lover*. Edmonton AB: NeWest Press.

Duffy, Michelle, 2013. The requirement of having a body. *Geographical Research*, 51, pp.130–36.

Hawkins, Harriet, 2011. Dialogues and doings: sketching the relationships between geography and art. *Geography Compass*, 5(7), pp.464–78.

Hawkins, Harriet, 2013. *For creative geographies: geography, visual arts and the making of worlds*. London: Routledge.

Hawkins, Harriet and Straughan, Elizabeth, eds., 2015. *Geographical aesthetics*. Surrey: Ashgate.

Hayes-Conroy, Allison, 2010. Feeling slow food: visceral fieldwork and empathetic research relations in the alternative food movement. *Geoforum*, 41, pp.734–42.

hooks, bell, 1989. *Talking back*. Boston: Southend Press.

Kraftl, Peter, 2013. *Geographies of alternative education*. Bristol: Policy Press.

Madge, Clare, Noxolo, Pat and Raghuram, Parvati, 2004. Bodily contours: geography, metaphor and pregnancy. In: Women and Geography Study Group, eds., *Geography and gender reconsidered*. London: Women and Geography Study Group. pp.68–83.

Madge, Clare, 2014. On the creative (re)turn to geography: poetry, politics and passion. *Area*, 46, pp.178–85.

Madge, Clare, 2016. Living through, living with and living on: creative cathartic methodologies, cancerous spaces and a politics of compassion. *Social and Cultural Geography*, 17(4), pp.207–32.

Marston, Sallie and de Leeuw, Sarah, 2013. Creativity and geography: toward a politicized intervention. *Geographical Review*, 103, pp.iii–xxvi.

Moss, Pamela, 2014. Some rhizomatic recollections of a feminist geographer: working toward an affirmative politics. *Gender, Place and Culture*, 21(7), pp.803–12.

Nash, Catherine, 1996. Reclaiming vision: looking at landscape and the body. *Gender, Place and Culture*, 3, pp.149–69.

Noxolo, Pat, Raghuram, Parvati and Madge, Clare, 2008. 'Geography is pregnant' and 'Geography's milk is flowing': metaphors for a postcolonial discipline? *Environment and Planning D: Society and Space*, 26(1), pp.146–68.

Pain, Rachel and Staeheli, Lynn, 2014. Introduction: intimacy-geopolitics and violence. *Area*, 46(4), pp.344–7.

Philo, Chris, 2014. Insecure bodies/selves: introduction to theme section. *Social and Cultural Geography*, 15, pp.284–90.

Pratt, Geraldine and Johnston, Caleb, 2014. Filipina domestic workers, violent insecurity, testimonial theatre and transnational ambivalence. *Area*, 46(4), pp.358–60.

Pratt, Geraldine and Rosner, Victoria, eds., 2012. *The global and the intimate: feminism in our time*. New York: Columbia University Press.

Price, Patricia, 2013. Race and ethnicity II: skin and other intimacies. *Progress in Human Geography*, 37(4), pp.578–86.

Richardson, Michael, 2015. Theatre as safe space? Performing intergenerational narratives with men of Irish descent. *Social and Cultural Geography*, 16(6), pp.615–33.

Sharp, Joanne, 2014. The violences of remembering. *Area*, 46(4), pp.357–8.

Sheridan, Dorothy, Street, Brian and Bloome, David, 2000. *Writing ourselves: mass-observation and literacy practices*. Cresskill, NJ: Hampton Press.

Smith, Sara, 2016. Intimacy and angst in the field. *Gender, Place and Culture*, 23(1), pp.134–46.

Tamas, Sophie, 2009. Writing and righting trauma: troubling the autoethnographic voice. *Forum Qualitative Sozialforschung*, 10(1), article 22. Available at: http://www.qualitative-research.net/index.php/fqs/article/view/1211/2642. [Accessed 9 November 2016].

Tamas, Sophie, 2014. Scared kitless: Scrapbooking spaces of trauma. *Emotion, Space and Society*, 10, pp.87–94.

Valentine, Gill, 2008. The ties that bind: towards geographies of intimacy. *Geography Compass*, 2(6), pp.2097–110.

Verhage, Florentien, 2014. Living with(out) borders: the intimacy of oppression. *Emotion, Space and Society*, 11, pp.96–105.

7 Writing/drawing experiences of silence and intimacy in fieldwork relationships

Kacy McKinney

How comics write stories

In her book, *Graphic women: life narrative and contemporary comics*, Chute (2010, p.2) notes the large and ever-expanding body of graphic narrative work – works by women that 'represent a new aesthetics emerging around self-representation', that are 'experimental and accessible' and that treat subjects relevant to feminist inquiry. In addition to widely read and acclaimed works such as Satrapi's *Persepolis* (2003) and Bechdel's *Fun Home* (2006), Chute's list of examples of graphic narrative works fills fourteen lines of text. Feminist works are integral to the wider history of comics, from underground comics of the 1970s and 1980s to the graphic narratives and memoirs of today (Chute, 2010). These comic forms are of particular interest to feminist scholarship as objects of study for their rich theoretical and empirical significance (Chute, 2010; Donovan, 2014), though scholars have been slow to recognize this. Extending beyond the analysis of the form and content of comics as a way to tell a story, I want to argue for the creation of comics and graphic narratives as approaches to writing feminist research, and that graphic narratives can be used both to write stories differently and to write different stories.

Writing comics and graphic narratives through the combined use of word and image offers a productive means to explore experiences of complex relationships in fieldwork. As an illustration of this approach, the graphic narrative I include below tells a story of silence and unexpected revelation while conducting research in rural India. The act of combining the writing and drawing of field research departs from conventional academic prose by opening up both literal and figurative spaces of encounter as a novel way to both *write* and *read* stories. I use the word *comics* throughout this chapter – despite the messiness and contested status of the term – to refer to text and image used together to tell a story in a way that neither could accomplish alone (McCloud, 1993; Abel and Madden, 2008; Eisner, 2008). I am also focusing on a specific subset of comics: *graphic narrative*, following Chute (2010), and *graphic memoir*, as described by Donovan (2014). In the graphic narratives and graphic memoirs discussed by these authors self-representation and speaking to embodied experiences and non-normative lives are central to the language of comics.

Graphic narratives have the capacity to create deeply intimate worlds for the author and the reader that can transport both to another place and time. Through an alternative use of space on the page and through the use of subtle visual language comic artists can set paces and alter perceptions of the passage of time in a story. With comics it is possible to amplify meanings through simplifying language and images (McCloud, 1993, p.30). Some of the most striking comics use the simplest imagery to convey experiences and ideas that relate more broadly to the human condition. In addition, the comic form has its own kind of intimacy. Through drawn images and handwritten text, comics can offer 'intriguing aesthetic intimacy' (Chute, 2010, p.6) that envelops the reader.

Comics are uniquely situated to address the expression and communication of intimacy in feminist geographical research. Comics can blur the boundary between author and audience. According to McCloud (1993, p.36), readers are invited not only to *look* at the comic, but also to *participate*. Comics can create openings for multiple interpretations and readings. For Chute (2010, p.3), the graphic narrative can provide an 'expanded *idiom of witness*, a manner of testifying that sets a visual language in motion with and against the verbal in order to embody individual and collective experience, to put contingent selves and histories into form'.

Silence, intimacy and researcher/research assistant relationships

As I prepared to do research, much like other feminist researchers, I tried to plan for everything ahead of time. I tried to think of everything. One of my many questions had to do with conducting fieldwork as a gay woman in an area of the world where homosexuality is criminalized. I spent time thinking about how I would perform my identity while in certain public and private spaces in India. For anyone doing research in spaces unfriendly to homosexuality, there are decisions to be made. When, if ever, should I reveal my sexuality? How should I negotiate this important aspect of my identity in my relationship with research assistants? Conducting fieldwork in a foreign country and foreign language(s) can mean that the relationship with a research assistant is of critical importance and that can present unexpected challenges. For researchers, working closely over long stretches of time and in spaces ranging from public to very private, research assistants can be the ones we grow closest to. Following Johnston (2010), I view this relationship as a site of secrecy and silence in research – one made all the more complex to navigate by its intimacy.

Feminist geographers have written extensively on the complex negotiations of power, privilege and ethics in fieldwork. In particular, many have noted the influence of the researcher's identity in interactions with research participants (for example England, 1994; Katz, 1994). Stiell and England (1997) highlight questions about the role of identity in their discussion of the differences and similarities between them as co-researchers. Ellis (2007) draws attention to relational ethics and responsibilities within friendships in research. Swanson (2008) discusses the ways in which different aspects of our identities can alter power dynamics in research, and how issues of power, privilege and vulnerability can surface in unexpected ways in the field. Relationships with research assistants, in particular, rely upon mutual trust, respect and understanding and as such also require careful negotiation (Hapke and Ayyankeril, 2001). Cupples (2002) argues for the need to engage our positionalities more fully, working to understand our performances of sexuality in fieldwork for a better understanding of the knowledge we produce.

As a gay woman doing field research in rural India I assumed that those around me would think I was heterosexual. Like Valentine (2002) I did not disrupt this assumption. Johnston (2010, p.293) highlights, 'the way in which secrecy and silence discourses about sexuality shift and change depending on the place of research', and for me this extends also to time. There are critical moments when the performance of identity in the field changes. I relate my own experience of some of these moments in the following two sections: one in written prose, the other in hand-drawn images.

Relating in the field

I had never kept my sexuality a secret; I had never needed to. It was 2008, and I was twenty-eight. I was in India doing my dissertation research on working children and agricultural change. I had hired a woman a few years younger to be my research assistant, and amazingly, ended up with the assistance of another woman interested in my research. We spent long days together, driving around in an old jeep on dirt roads and in dry riverbeds. We were a sight to behold in the hills of Rajasthan, three young women – two Indian and one American. We were all dressed in different varieties of Indian clothing, though not all pulling it off. A local tribal man also accompanied us; he generously drove us around, negotiating intoxicated men and translating difficult words.

Over several months, these two women and I got to know each other. We joked and laughed and observed and discussed. They wanted to know what I was looking for. They wanted to discuss the questions I was asking and why I was asking them. They wanted to see what the process of research I had been trained in involved. They encouraged me, and they helped to expand my under-standing of what we were seeing and of the context for it. They did not simply *assist* me, they guided and listened and explained. We stayed together and spent early mornings and late evenings cooking, talking and having fun. Our conversa-tions turned to any number of topics and we shared stories and memories with each other. They helped to shape my research and we became friends.

During that time I decided not to talk with them about my partner. I struggled with this decision. Was it more important to maintain the rapport we had built or to be open with my collaborators about my identity? This working relationship was of such vital importance to me that I felt I could not take the risk of a negative response. So I stayed silent. Everywhere we went women asked about our mari-tal status and whether we had children. I easily evaded the question for the most part. The women we interviewed were more interested in the two Indian women in that regard, perhaps because they could relate to them more easily. However, when pressed, I referred to my *husband*, rather than to erase her existence completely, even though it played havoc with her subjectivity. On the rare occasion that I uttered the word *husband* it rang so false in my ears; I never spoke of this in the private spaces with my research assistants, where I managed to avoid the subject altogether.

In the graphic narrative below, I share an unexpectedly humorous process of revealing my sexuality to these two women. In the creation of this piece, I sought to highlight the awkwardness of the negotiation of intimacy and subjectivity in researcher/research assistant relationships and the powerful significance of private spaces in the field for silences, secrecy and revelation. I am seeking to share with the reader an intimate view into my flawed process of determining where to place my trust and how I judged which people could accept my sexuality. In the end, I had only confused my friends with my occasional mention of a *husband*. For as one of the women exclaims at the end of the graphic narrative, now it made so much sense! There was something I was not saying – something had seemed *off* to her: incomplete, held back. The moment, or rather the space, of *coming out* portrayed in this piece relates both to the need to understand the performance of identity in the field (Pratt, 2000), and to find new means to approach writing intimacy.

Coming out in the field

Figure 7.1 After a long day of interviews

Author: Kacy McKinney

Figure 7.2 We launch into a marathon of Almodovar films
Author: Kacy McKinney

Figure 7.3 The lights came back on

Author: Kacy McKinney

Figure 7.4 I start to feel a conversation coming

Author: Kacy McKinney

Figure 7.5 At first, silence

Author: Kacy McKinney

Figure 7.6 I'm gay

Author: Kacy McKinney

Comics as a means to write research

The graphic narrative above demonstrates how this form of expression can be used as a means to write research experiences. As feminist scholars have long noted fieldwork brings with it all manner of challenges, dilemmas and nego- tiations as we seek ethical, reflexive and transformative praxis. The building of relationships during fieldwork presents a particularly complex terrain of power, emotion and intimacy. As feminist scholars, we need more – and different kinds of – spaces for writing about these experiences. This graphic narrative helps me alter the pace of my story as well as shift the audience experience. It also affords room for multiple interpretations, understandings and resonances. Comics pre- sent exciting potential both as objects of study and as method of representing and reporting on research. Within comics time is represented as space (Chute, 2010, p.7), and the spatial relationship to the page is altered for both author and reader. This transformation of the page provides a critical opening for expres- sions of intimacy in field research. Like Chute (2010) and Donovan (2014) I am anxious to see comics – particularly works by women and those representing non-normative lives – taken seriously for their contributions on theoretical and empirical grounds.

Works cited

Abel, Jessica and Madden, Matt, 2008. *Mastering comics: drawing words and writing pictures continued.* New York: First Second Books.

Bechdel, Alison, 2006. *Fun home: a family tragicomic.* New York: Houghton Mifflin Harcourt.

Chute, Hillary L., 2010. *Graphic women: life narrative and contemporary comics.* New York: Columbia University Press.

Cupples, Julie, 2002. The field as landscape of desire: sex and sexuality in geographical fieldwork. *Area,* 34(4), pp.382–90.

Donovan, Courtney, 2014. Representations of health, embodiment, and experience in graphic memoir. *Configurations,* 22(2), pp.237–53.

Eisner, Will, 2008. *Comics and sequential art.* New York: W. W. Norton.

Ellis, Carolyn, 2007. Telling secrets, revealing lives: relational ethics in research with inti- mate others. *Qualitative Inquiry,* 13(1), pp.3–29.

England, Kim, 1994. Getting personal: reflexivity, positionality, and feminist research. *The Professional* Geographer, 46(1), pp.80–89.

Hapke, Holly M. and Ayyankeril, Devan, 2001. Of 'loose' women and 'guides', or rela- tionships in the field. *The Geographical Review,* 91(1–2), pp.342–52.

Johnston, Lynda, 2010. The place of secrets, silences and sexualities in the research pro- cess. In: Róisín Ryan-Flood and Rosalind Gill, eds., *Secrecy and silence in the research process: feminist reflections.* Routledge: New York, pp.291–305.

Katz, Cindi, 1994. Playing the Field: Questions of fieldwork and geography. *Professional Geographer.* 46(1), pp.67–72.

McCloud, Scott, 1993. *Understanding comics: the invisible art.* New York: Harper Collins.

Pratt, Geraldine, 2000. Research performances. *Environment and Planning D: Society and Space*, 18(5), pp.639–51.

Satrapi, Marjane, 2003. *Persepolis*. New York: Pantheon.

Stiell, Bernadette and England, Kim, 1997. Domestic distinctions: constructing difference among paid domestic workers in Toronto. *Gender, Place and Culture*, 4(3), pp.339–59.

Swanson, Kate, 2008. Witches, children and Kiva-the-research-dog: striking problems encountered in the field. *Area*, 40(1), pp.55–64.

Valentine, Gill, 2002. People like us: Negotiating sameness and difference in the research process. In: Pamela Moss, ed., *Feminist geography in practice: research and methods*. Oxford: Blackwell, pp.116–26.

8 Open for business?

First forays into collaborative autobiographical writing in extractive northern British Columbia

Zoë A. Meletis and Blake Hawkins

Foreword

As feminist geographers, we value including local voices in considerations of topics such as extractive economies. We have both lived in northern British Columbia (BC) for a number of years. Northern BC is a vast area of roughly 600,000 square kilometres, a size similar to that of France (Northern Health Authority, 2015), and approximately 300,000 residents. We recognize the emotionality associated with living in places under frequent promise/threat of additional extractive development (mining; oil and gas; forestry). We conceived of writing this chapter as an opportunity to reflect on our own communities and surroundings. We hope that our writing does justice to the intimacy we have felt while living in northern BC. Searching for a way to write intimate relations into texts on extractive economies and pressures, we experimented with autobiographical writing and we share extracted syntheses from our greater stories here. Zoë wrote from Prince George (PG) and Blake from Kitimat, BC. We present these writings here, offering personal accounts of life under extractive pressures, in northern BC. Sharing stories from our lives allows us to reflect upon our first forays into intimate writing and what they have taught us about the places in which we live, ourselves and autobiographical writing.

Our first forays: Autoethnography-cum-autobiographical writing

This project began with casual discussions about how concepts in our research (space, place, place-connectedness and identity) seemed to feature in uncomfortable experiences we had undergone or witnessed in northern BC. Originally, we sought to write autoethnographic components on life in two extractive economy locales and to join them here. We understood autoethnography as an approach to research and writing that seeks to describe and analyse personal experiences within a broader cultural context (Ellis, Adams and Bochner, 2011), while offering analytical power and empirical richness (Chang, Ngunjiri and Hernandez, 2012). Personal stories and discussions of shared patterns between them seemed to hold great potential for micro-level perspectives on larger scale phenomena in extractive northern BC.

Our work became more autobiographical than autoethnographic, however. We were emphasizing *our* experiences within Prince George and Kitimat, rather than including our own stories among other ethnographies on greater patterns and themes. Our conceptualization of autobiographical writing is similar to Purcell's (2009, p.234): 'a method of inquiry in which the research narrates his or her own life. It shares much with biography, ethnography and autoethnography in that they all aim to provide a rich account of human experience.'

Recounting personal experiences, we situate these within our understandings of PG and Kitimat. We also reflect on our privileged positions and our understandings of how social inequality and marginalization could be exacerbated by change or the promise of change. Zoë emphasizes contrasts between the lives and roles people in PG may envision for themselves, in contrast with the realities of local relationships with extraction. Blake describes how he witnessed and experienced extractive/economic/social changes via family life and friends' accounts in Kitimat.

Understanding our work to be autobiographical rather than autoethnographic afforded us more freedom to write up our experiment and to choose which lessons to share from it (Purcell, 2009). Our confused attempt at autoethnography seemed in fact to be a joint exercise in phenomenological autobiography. This is a form of autobiography that describes everyday life experiences, from a personal standpoint. It emphasizes that there is significance to what one experiences in one's everyday life. This spoke precisely to what we had hoped to do by sharing perspectives from our lives. We wanted to help humanize the ways we think about experiencing resource extraction and to personalize the impacts of living with extraction, the threat of future extraction and the impacts associated with boom and bust extractive economic cycles.

Despite some advocates (e.g. Valentine, 1998; Moss, 2001; Besio and Butz, 2004), human geography does not yet include a prevailing culture of acceptance for autoethnographic and autobiographical research (Longhurst, 2012). Using the personal or individual point of view as key information is still largely viewed as a radical departure from standard research methods (Butz and Besio, 2004). This may help to explain why we could not easily find any first person-focused accounts of life with extraction in Canada. This is a problematic gap in the literature since life within communities with strong ties to extraction, both existing and proposed, is deeply personal, although also tied to largely social forces and phenomena. It is experienced through brain and body and through a person's relationships with land and resources. It was these very personal connections, experiences and networks related to place, identity and emotionality that we had hoped to highlight elucidate via writing our own experiences.

Chapter structure and rationale

The next section of this chapter is composed of our two autobiographical texts. These are followed by reflections on these and lessons learnt along the way. The reader will see that while we do reference greater patterns and communal concerns as we saw them, our stories focus on key experiences and exchanges that we had as individuals. We hope that sharing our personal stories emphasizes the heavy reliance on extractive economies the emotionality of life in northern BC. We wanted to give readers an idea of the daily realizations, discomforts, upsets, disappointments, hopes, fears and concerns that come from living in such places, so that they might consider these alongside claims and arguments about community-level impacts (which seem to dominate the related literature). In the end, we met our core goal, despite deviations in our methodological course, and we offer our autobiographical writing accounts as examples of how such texts can elucidate impacts of extractive economies, at the personal level.

Autobiographical texts

Environmental attachments, concerns and imaginings in Prince George and surroundings: Zoë

Prince George sits at the intersection of the Fraser and Nechako rivers, on traditional Lheidli T'enneh territory. Not built for any particular industry, it has had various industry-focused incarnations. It was a fur trading station in 1915 and forestry was the primary employer from the 1950s until the 1990s (Halseth, 1998). In the 1990s, mountain pine beetle damage and softwood lumber disputes decimated the regional forestry industry (Hanlon and Halseth, 2005), which contributed to an economic slowdown (Halseth, Sedgwick and Ofori-Amoah, 2007). The past five to ten years have brought economic revitalization, with the University of Northern British Columbia, the University Hospital of Northern BC, a regional health authority and smaller industries (Northern Development Initiative Trust, 2015). Despite diversification, the region remains reliant on extractive industries and associated investments and infrastructure (e.g. Prince George is a shipping and transportation hub).

A transplant to the area, I am impressed by regional pride of place and place-connectedness; residents speak passionately about PG. Also, many people are literally tied to land, water and other northern resources via jobs or residences close to resources and/or exploitation. Land and resources are key currents running through people's lives, hearts, households and communities.

In contemplating intimate connections to land and resources as well as the emotional aspects of life with extraction, I identified three key relationships/experiences in my PG life: (1) my connections with a documentary film called *Line in the Sand* (2015); (2) my participation in the Joint Review Panel (JRP) of the Northern Gateway project (summer 2012); and (3) my involvement with the Prince George hosted Canada Winter Games (2015).

In 2014, I contributed to the film *Line in the Sand* (2015), which traces the proposed route of the Enbridge Northern Gateway Pipeline project, speaking with diverse actors along the route. The filmmakers came to my environmental justice class, for interviews. At that time, among other things, we were being bombarded with corporate messaging about what extraction could do for our northern communities and regional employment (e.g. pamphlets in mailboxes; ads before movies screenings). Often, I felt violated and annoyed by corporate, government and part-nership messaging in PG. Uninvited, it crept into social media feeds and arrived at our doorsteps. We were witnessing apparently polarized debates in the media, in our classrooms and on the streets of PG. These seemed to reinforce the false dichotomy of jobs vs environment, forcing uncomfortable and divisive choices. Both national and provincial governments at the time were also pro-extraction. I took comfort in the surprising unity that I saw and in actions of resistance. I noted how diverse opposition to extraction including the proposed pipeline was and how activists were getting along despite differences.

As a local resident and a scholar of environmental justice and political ecology, I felt compelled to participate in the film and the Prince George Community Hearings of the Joint Review Panel (JRP) on the proposed Enbridge pipeline. Powerful extractive proponents made resistance seem nec-essary but dangerous, especially with growing mainstream discourses about Canadian authorities watching environmental activism as a type of domestic threat. Colleagues and I joked about being on a list somewhere but, really, the prospect was unnerving.

In addition to thinking about risk, I was also thinking and writing about local fears and concerns. The oppressive pro-extraction atmosphere permeated my classes on the politics, process and power dynamics of environmental decision-making. Case studies from our region brought issues home in new and unpleasant ways. Students asked difficult questions about existing and proposed develop-ments and I had no good answers for them.

In my PG experiences, threats to spaces and resources within them seem inextricable from threats to place and identity. This was evident when I partic-ipated in the Community Hearing of JRP on the proposed Enbridge Northern Gateway Pipeline on 09 July 2012 (Hearing Order OH–4–2011). It was an indel-ible experience. Common threads spanned across oral statements from diverse participants. People spoke of how the proposed pipeline would change/threaten spaces, resources, populations and ways of life. Participants questioned whether they could still consider themselves good stewards of the land if their actions did not prevent the pipeline. It was a distinctly emotional set of testimonies. Most people in the room sounded deeply invested in these places, well beyond eco-nomic value and existing or potential employment. Participating in the JRP was an oddly unifying, community-building experience. I feel bonded to the people that spoke during our allotted PG community days. I also witnessed much support in the room, across all demographic categories. I did not expect it to be such a multi-faceted emotional and learning experience.

Another key experience was the PG-hosted Canada Winter Games in February 2015. The multiyear lead-up to the Games was replete with pro-extraction cor-porate messaging. The Games were decorated with corporate logos and included high level or legacy extractive sponsors. Companies had also donated staff hours and/or equipment use, making them more visible than usual in PG. Such contri-butions were hard to argue against – they were funding our successful bid, our efforts to be good hosts and Canadian athletes.

As a volunteer and citizen participant, the Games were enjoyable but taxing. We were called upon to celebrate PG, and yet we were also escaping everyday PG life via a temporary series of events and festive atmosphere. This pushed us to consider our realities in PG and also invited us to think about how different life and economy in PG *could* be. For example, the Games slogan was 'Illuminate the North.' This led me to consider which industries illuminate the north, both figuratively and literally. I also wondered how many local and regional identities and livelihoods end up being inextricably linked to industries that do not prioritize sustainable planning for our future. The Games was a dual demonstration of our wonderful community and our disturbing ties to corporate extraction – all of this as we dealt with pressure to open up more of the north to oil, gas, natural gas and pipelines. We had seen our downtown streets filled with people participating in a different economy (tourism, sporting events, leisure). Games marketing also encouraged us to imagine Prince George as a different place, by emphasizing our unique landscapes alongside our social and cultural capital. This was hopeful but also disturbing given our real everyday dependencies and pro-extraction leaders. Again, it was a surprisingly emotional and reflexive experience, particularly when writing it up.

A Window into Life and Changes in Kitimat, BC: Blake

I lived in Kitimat for most of my life, until the fall of 2014. Now based in Vancouver, I come and go for long stretches of time. To many, Kitimat is an industry town in northern BC. To me, Kitimat is home – a place that has seen many projects and promises cycle through, to the benefit and detriment of local peoples, cultures and landscapes.

When we began this project, both Kitimat and PG were being portrayed as booming, with real estate markets heating up based on existing and prospective extractive and extraction-related projects in the region (McCreary and Milligan, 2014). As we drew this chapter to a close, rental housing remained hard to come by in both places and the low Canadian dollar and waning interest in Canadian oil and gas were lessening some development pressures on each. Simultaneously, though, the extractive cores of each economy (tied to existing and prospective extraction) are currently helping to buffer against recessionary forces.

Kitimat is an industry and unionist town. As a settler community, it includes Portuguese, Indo-Canadian, Fijian and Finnish influences. It is a town created by the aluminum company ALCAN (now Rio Tinto ALCAN) to support an aluminum smelter, in the 1950s. Initially planned for 35,000 to 50,000, its population has never reached this great size. Typically, its population lies between 9,000 to 15,000 people. During the 1950s, the BC provincial government followed a Fordist economic model (Larsen, 2004; 2008; Markey, Halseth and Manson, 2009). This resulted in support for big industry and a government emphasis on an extractive economy derived from exploiting natural resources in central and northern BC. Kitimat became a boomtown and drew migration (Herkes, 2010). Since the initial boom, there have been some periods of economic hardship, mainly in the 1990s and 2000s, due to the closure of two major employers (Methanex and Eurocan).

As an undergraduate student at the University of Northern British Columbia, I had a feminist geographer mentor. I came to afford greater value to the insights contained in individual personal stories from small town community members and learnt about arguments for their academic merit. Later, I took a feminist research methods course that motivated me to experiment with intimate qualitative research, which I saw as having great potential power. I was inspired to attempt an autoethnography of my life in Kitimat during the summers of 2013 and 2014, for this chapter.

Discussions of 101 pipelines and extraction in northern BC were part of my family's discussions long before we started this project. For example, they frequently informed me about ongoing developments related to the Rio Tinto Alcan modernization project (2012–2015). When I began writing about Kitimat, I spent a lot of time thinking about recent changes in the community. Some of the development could be interpreted as positively impacting our community. For example, as the result of increased corporate and especially extractive interest and investment, we gained local infrastructure and service improvements that benefitted more than those directly employed by Alcan. Also, the associated arrival of new stores and restaurants are often exciting side benefits in a town with few such options. I witnessed and partook in local enjoyment of such new additions in 2013 and 2014. This, however, was contrasted with gaps in necessary services and goods. I encountered people struggling to satisfy basic needs like food and housing. I also saw residents affected negatively as I learnt about resentment of unwelcome changes.

It struck me again that despite associated (re)investments and profit generation, no combination of development project(s), past or present, ever seemed to bring widespread relief for citizens struggling to make ends meet. For example, prices of consumer goods and food remain high, partly due to the small population size and its relatively remote location. Such hardships can be especially hard to bear since despite its distance from major centres, Kitimat is not far north enough for its residents to be considered as northern Canada and thereby qualifying for federal cost of living subsidies. When I travelled home, I was often shocked at how much more expensive groceries are in Kitimat than in Prince George, the nearest larger urban hub. Discussing such price differences with family members often causes friction and raises questions about the true benefits of development in Kitimat. In the last few years, I have also noticed more and more residents going to the food bank. Further, my family personally experienced difficulties associated with lags between extractive booms and poor social service catch-up to meet population increases and related social issues.

While writing about my life in Kitimat for this project, I found comments indicative of local challenges appearing on social media platforms that I use. Facebook pages were captivating, telling suggestions about everyday life in Kitimat. On the Facebook page *Kitimat Politics*, for example, some people share everyday concerns about the future of the town and raise questions and criticisms of elements of its development profile. Comments here suggested that residents were upset about increasing costs of everyday essentials (e.g. groceries) and their tenuous ability to maintain long-term access to reliable, accessible food sources.

Most such comments from community members placed the blame for all such problems on increases in resource development, often without elaborating on how these relationships play out. While carrying out our project, I was surprised by the apparent lack of community support for any of the (new) projects; I did not see a single supportive comment. Instead, I read countless posts about the town being ruined by new industries and projects. Comments often suggested that the town was no longer a peaceful place to live. For example, a powerful local discourse about the presence of unfamiliar or so-called transient people as being different from longer term Kitimat residents existed across some comments.

To me, residents seemed apprehensive about the rapid pace of development. Their negative comments and concerns, as I saw them, reflected greater tensions between community wants and needs, and those of corporate extractive development. I also noticed that many of those commenting about such concerns listed employment positions with the very industries they were dissatisfied with. The contrast between the struggles of everyday life in Kitimat and peoples' perceptions of these seemed in great contrast to prevalent corporate and government development-related promises and narratives in northern BC.

Making sense of our experimenting with autobiographical writing

We learnt a great deal throughout this project and share five main reflections here.

1 Autobiographical texts can contribute micro-level insights into life with extraction, in northern BC

Initially, we saw intimate writing as a compelling way to include local voices and place-based vantage points in contemplations of places where cycles of extractive promises and resulting impacts abound (including those of abandoned projects). We also viewed it as a tool for exposing larger patterns of emotions and experiences across individual experiences, when combined with scholarly contextualization. We began this project because we recognized we were *already*, as part of our everyday lives, experiencing and witnessing behaviours, reflections, emotions and insights that should be being shared with larger audiences to personalize and provide micro-level insights into life under extractive pressure. The literature tends to discuss resource economy impacts in northern BC region at the community scale (Larsen, 2004, 2006, 2008; McCreary and Milligan, 2014). We hope that our autobiographical accounts included here act as compelling examples of the potential for incorporating intimate texts into considerations of human–environment interactions, as complements to less personal, greater scale analyses. Our aim was to provide examples of intimate details about how people can be impacted via psychological impacts and impacts on peoples' relationships.

We wanted to offer glimpses into types of impacts typically less obvious to outsiders and rendered somewhat invisible in community-scale analyses.

2 Autobiographical writing is incompatible with busy lives

In addition to struggling with the messiness of our own lives and the concept of recording and sharing emotional experiences, we also struggled to maintain disciplined observational and recording practices. Life is messy and disorganized. For this reason, we found it harder than expected to record our own lives and observations of these in a consistent way. Further, life does not unwind in a linear and logical way, with immediate realization of the significance of one's observations shining through at first encounter. Thus, we struggled with adequately capturing our lives and our reflections upon them, as related to extraction and the threat of further such development in the north. Sometimes, for example, Zoë would only experience anger over extractive messaging after experiencing invasive advertising (e.g. in social media feeds; in the movie theatre as pre-move ads; in local newspapers; on the radio) over the course of months. How best then to describe how this anger grew and swelled, when it began and what it represents? We are still not quite sure.

In reflecting upon our inconsistent process, we wonder if we should have developed a more rigorous material practice that would have facilitated consistent recording and encouraged regular reflection. We found that a lot of our most insightful reflections were quite fleeting because they tended to occur at academically inconvenient times like when we were out for a run, taking a shower or engaged in a fascinating encounter far from our notebooks and laptops. Perhaps we should have made greater efforts to be at-the-ready for note taking and analysis. But this raises the challenge of trying to live life authentically while simultaneously recording it diligently. We believe that more consistent note taking may have in fact disrupted some of our best reflections on the very themes we sought to engage with. As autobiographical novices, we have not figured out how to best reconcile these yet.

3 Experiencing a bumpy research process taught us a great deal

Throughout our process, but particularly when writing this chapter, we struggled with how best to craft the bumpy process and resulting notes into a cohesive text. First, the logistics of process created challenges. To start, we veered a good distance from our original plan. We also found that our individual and variable research and annotation processes made our data challenging to analyse and to write up. We collected materials and notes in different ways. Second, it is not easy to record, examine and put one's own life under a microscope. We had very productive discussions with each other, but then struggled to integrate these with our written accounts and to translate our experiences into relevant knowledge for others. Third, we began to ask higher-order questions about the worth, value and spirit of autobiographical writing, and to wonder what understanding could be gained by creating more space for it in human geography scholarship. Lastly, as the project evolved, we gained gratitude for leaving our comfort zone and standard research practices, and for the work our more seasoned autobiographical researcher colleagues contribute. Van Maanen (2004, p.444) suggests that ethnographers are at 'their best when mucking around the empirical base camps of social science than when perched (always precariously) on some theoretical mountaintop'. This is how we felt about this endeavour. We successfully mucked about, and learnt a great deal about ourselves and our research practices, meeting the spirit of our intimate writing attempt.

4 We were poorly trained for this undertaking and our academic educations had cautioned us against intimate writing

In reflecting on our mucking about in this research and writing, we realized that we both felt ill-equipped to write up our autobiographical data. For one, our training had generally favoured distancing from research subjects, despite our largely qualitative expertise. Also, we sometimes felt that we lacked the right vocabulary or prose to tell our stories with ease. While it may be true that 'ideas about empirical evidence, objectivity, reason, truth, coherence, validity, measurement and fact no longer provide great comfort or direction' (Van Maanen, 2004, p.435), such powerful influences still permeate formal academic education. During our degrees, we had both been deeply steeped in such influences, despite our wanting to escape them. We came with no coherent answer to questions such as: how do you write down your innermost feelings and thoughts in a somewhat detached academic style? Is this possible? Why do we feel that this is necessary?

Knowing your own life and stories within it is one thing, but finding a compelling way to share them, especially within the confines of academic writing, is quite a different matter. Denzin (2004, p.449) describes 'moving from the field to the text to the reader' as a 'complex reflexive process'. This was especially true when dealing with the field inside of us. We believed that there was unique value in sharing our individual and common experiences as residents living with extractive pressure in northern BC, however we struggled with converting them into a more formal format while also trying to ensure accessibility. We had a sense of common themes and key terms that could be used to describe patterns we were observing and experiencing and yet we found them hard to articulate and explain, especially to more distant readers. For example, we observed and were involved with much talking and writing about pride of place in northern BC, although not necessarily using that term. We found it hard, however, to effectively convey the *exact* nature of people's concerns when restricted to words on a page. Sometimes, our proximity to issues made them harder, not easier to explain to distant readers because we knew what was going on, and could talk about it with each other, but found it hard to find the right words to explain phenomena to others when restricted to words on a page, without intimate knowledge and access to local ambiance as we felt it and understood it. Like Rosaldo (1989), we worried about reducing our experiences to caricatures and found it hard to share our micro-geographies authentically.

5 Intimate research and writing demands deep personal investments, uncomfortable revelations and thought processes, and the courage to share these

In human geography, it is typical for authors to temper emotionality when writing. Given this, it felt unnatural and overly revelatory of ourselves to consider how we could best share our observations and our feelings about these. It felt risky to offer our own lives and experiences up for scrutiny. We were also unsure of how to properly set up the stories that we witnessed. We did not know how much our audience would care and we were not sure what they would like to know. As two geographers in the north, we felt particularly paralysed by our inability to adequately describe related senses of place and how they were being threatened. Also, while we both often pay great attention to ensuring that respondent quotes are recorded and used verbatim in our other work, we felt pressure to reword our own notes and observations in order to bring them up to an imagined correct academic level.

Further, it was hard to contemplate sharing our most intimate wranglings with threats to our own sense of place when writing about home – a place of utmost importance. For Zoë, the most powerful manifestation of this was her experience when participating in the JRP Prince George Community Hearing, on 9 July 2012. She remains surprised at how the emotional stakes seemed to change when it was her turn to speak. Zoë only realized how passionate she was about what she had written (not wanting to teach about the pipeline and likely negative outcomes) when the words began to come out of her mouth. Zoë still has trouble speaking to the statement she gave that day, even though all contributions are on the public record (e.g. National Energy Board, 2012). These citizen statements include the powerful stuff that composes autobiographical stories about place but were documented in an impersonal government compendium format. Place moves people, it inspires people and it pushes them to great emotion. And all of this is why writing the intimate is risky and difficult, especially when the stakes are high, as they tend to be in an atmosphere of looming resource extraction domination and academic careers situated within such precarious economies.

Closing thoughts on intimate writing and geography

Our methodological experimentation left us with many questions about autobiographical writing, in terms of process, the spirit of autobiographical writing and how best to incorporate it into human geography of the environment. As we discussed shortcomings with our research process, for example, we began to wonder if researchers demand of it a higher documentation standard than others forms of qualitative research, when possible and when appropriate: for example, when we undertake other forms of qualitative fieldwork, whether studying ecotourism and conservation in Costa Rica (Zoë) or LGBTQ youth and health in northern BC (Blake). It is also generally accepted that researchers must be flexible and adapt to the needs of their research project and, perhaps most importantly, the needs of participants. We also seem to take it for granted that scholarly understanding of places, topics, peoples and phenomena will change over time. Considering all of this, why do we not make appropriate allowances for autobiographical and other forms of intimate writing? Do we hold stories or data about self to higher and perhaps less attainable standards than other forms of qualitative research? If so, what do we stand to lose?

As we draw this rich yet flawed collaboration to a close, we recognize the need to invest further in developing our understandings of intimate writing processes, practices and prose. One could easily argue that we were unsuccessful in that we deviated from our original course and scheduled topics for we had trouble both undertaking autobiographical research and writing it up. In the end, however, we deem this a successful project *because* of our great learning from the bumpiness. During this project, we struggled with articulating what we observed, experienced and simply knew by living. We worried about the consequences of revealing too much of our own lives and those of others. In these ways, we truly met the spirit of an intimate research and writing project. Unfortunately, we did not illuminate much about living with extractive pressures in northern BC and for this we are remiss. We hope, however, that the reader has enjoyed our candid descriptions of the uncomfortable journeys we took via our first forays into intimate writing. We would encourage others to push beyond their conventional or comfortable research practices, since we grew by leaving our typical approaches and methods behind for this project. Grateful for the learning we have gained via contributing to this book, we have also gained a deeper respect for the scholars who have contributed more eloquent chapters alongside ours. We look forward to gleaning their advice on accepting and respecting the bumpiness, while successfully crafting personal experiences into information that is all at once intimate, scholarly and accessible.

Works cited

Besio, Kathryn, and Butz, David, 2004. Autoethnography: a limited endorsement. *Professional Geographer*, 56(3), pp.432–38.

Butz, David, and Besio, Kathryn, 2004. The value of autoethnography for field research in transcultural settings. *Professional Geographer*, 56(3), pp.350–60.

Chang, Heewon, Ngunjiri, Faith Wambura and Hernandez, Kathy-Ann C., 2012. *Collaborative autoethnography*. New York: Routledge.

Denzin, Norman K., 2004. The art and politics of interpretation. In: Sharlene Nagy Hesse-Biber and Patricia Leavy, eds., *Approaches to qualitative research*. Oxford: Oxford University Press. pp.447–72.

Ellis, Carolyn, Adams, Tony E. and Bochner, Arthur P., 2011. Autoethnography: an overview. *Forum: Qualitative Social Research*, [e-journal] 12(1), Article 10. Available at http://www.qualitative-research.net/index.php/fqs/article/view/1589 [Accessed 14 July 2016].

Halseth, Greg, 1998. *Prince George: a social geography of British Columbia's 'northern capital'*. Prince George, BC: University of Northern British Columbia Press.

Halseth, Greg, Sedgwick, Kent and Ofori-Amoah, Benjamin, 2007. From frontier outpost to 'northern capital': the growth and functional transformation of Prince George, BC,

Canada. In: Benjamin Ofori-Amoah, ed., *Beyond the metropolis: urban geography as if small cities mattered*, New York: University Press of America. pp.18–42.

Hanlon, Neil and Halseth, Greg, 2005. The greying of resource communities in northern BC: implications for health delivery in already under-serviced communities. *The Canadian Geographer*, 49(1), pp.1–24.

Herkes, Jennifer. 2010. Planning for resilience: a case study of Kitimat, BC. MA. University of Northern British Columbia.

Larsen, Soren C., 2004. Place identity in a resource-dependent area of northern British Columbia. *Annals of the Association of American Geographers*, 94(4), pp.944–60.

Larsen, Soren C., 2006. The future's past: politics of time and territory among Dakelh First Nations of British Columbia. *Geografiska Annaler*, 88B(3), pp.311–21.

Larsen, Soren C., 2008. Place making, grassroots organizing, and rural protest: a case study of Anahim Lake, British Columbia. *Journal of Rural Studies*, 24(2), pp.172–81.

Line in the Sand, 2015. [video] Directed by Tomas Borsa and Jean-Philippe Marquis. Vancouver: Video Out.

Longhurst, Robyn, 2012. Becoming smaller: autobiographical spaces of weight loss. *Antipode*, 44(3), pp.871–88.

Markey, Sean, Halseth, Greg and Manson, Don, 2009. Contradictions in hinterland development: challenging the local development ideal in northern British Columbia. *Community Development Journal*, 44(2), pp.209–29.

McCreary, Tyler and Milligan, Richard A., 2014. Pipelines, permits, and protests: Carrier Sekani encounters with the Enbridge Northern Gateway Project. *Cultural Geographies*, 21(1), pp.115–29.

Moss, Pamela, ed., 2001. *Placing autobiography in geography*. Syracuse: Syracuse University Press.

National Energy Board, 2012. Enbridge Northern Gateway Project Joint Review Panel. Ontario: Canadian Environmental Assessment Agency. [website] Available at http://gatewaypanel.review-examen.gc.ca/clf-nsi/bts/jntrvwpnl-eng.html [Accessed 14 July 2016].

Northern Development Initiative Trust [NDIT] 2015. Northern success stories. [website] Available at http://www.northerndevelopment.bc.ca/explore–our–region/search/ [Accessed on 14 July 2016].

Northern Health Authority, 2015. About us. [website] Available at https://northernhealth.ca/AboutUs.aspx [Accessed 12 January 2016].

Purcell, Mark, 2009. Autobiography. In: Rob Kitchin and Nigel Thrift, eds., *International encyclopedia of human geography*. Oxford: Elsevier. pp.234–9.

Rosaldo, Renato, 1989. *Culture and truth: the remaking of social analysis*. Boston: Beacon.

Valentine, Gill, 1998. 'Sticks and stones may break my bones': A personal geography of harassment. *Antipode*, 30(4), pp.305–32.

van Maanen, John, 2004. An end to innocence: the ethnography of ethnography. In Sharlene Nagy Hesse-Biber and Patricia Leavy, eds., *Approaches to qualitative research*. Oxford: Oxford University Press, pp.427–46.

9 Walking the line between the professional and personal

Using autobiography in invisible disability research

Toni Alexander

In the fall term of 2013, I found myself applying for new faculty positions because of a variety of intersecting issues ranging from dual career concerns between me and my partner to an overwhelming lack of fit within my university department and college. I was the cultural geographer in the Department of Geology and Geography within the College of Sciences and Mathematics. At the time, I had held tenure for three years and could have remained indefinitely, but I finally came to the conclusion that my circumstances were not likely to change no matter how long I remained. I only applied to two positions in very specific programmes that seemed to address the problems I had encountered; both were chair positions within multidisciplinary departments (Global Cultures and Languages in the College of Liberal Arts and History and Geography in a College of Arts and Sciences). Having chaired multiple faculty search committees, I was fully aware that at some point in the process I would be asked to complete Equal Employment Opportunity Commission forms by those institutions. Logically, I knew that the search committees and deliberating deans themselves would not see the documents; but I still struggled with one question. Am I disabled? What may seem to be an easy question to most still plagues me even today as my answer may depend upon time and place. Over twenty years ago, I was diagnosed with depression and an eating disorder – conditions that continue to play a role in my life even today – however, with respect to employment or even my education, I've never requested any type of accommodation for this condition. If I've never disclosed my condition, does it mean I'm not disabled in the workplace? What has kept me from formally registering with my employer? Is it that I'm not sure what kind of accommodation might be made or that there are none possible? Or is it that the stigma attached to being labelled as mentally ill or disabled seems a threat to my career in higher education?

In this chapter, I investigate invisible disability within the context of the higher education environment. I use an autobiographic approach to explore the implications of personal mental health disorders for university faculty whose professional reputations and livelihoods exist in a place where intellect and reason are deemed to reign supreme. I explore the difficulties of intertwining the personal and professional despite the often proclaimed liberating elements of doing so.

Invisible disability

Disability is contemporarily regarded as a social response to an impairment whether it be physical or psychological (Dear, et al., 1997; Laurier and Parr, 2000; Braddock and Parish, 2001). As a social construction, disability is often defined in terms of social and/or physical surroundings. Geography, in particular, is well equipped to exploring those people (i.e. the disabled) who do not seem to meet the normalized expectations of place or disability (Hansen and Philo, 2009).

Scholars often make a distinction between visible and invisible disabilities (Tidwell, 2004; Moss and Teghtsoonian, 2008). Visible disabilities generally refer to those which are readily apparent to an observer and are often associated with biophysical impairments – at times made evident by the use of assistive devices. In contrast, invisible disabilities are those that are not always immediately discernible and may require regular interaction with the disabled person or the personal disclosure of their condition for others to be aware of it (Davis, 2005). Invisible disability occupies a far wider range of conditions, including physical impairments (diabetes, asthma, chronic pain etc.); sensory impairments (auditory and aural, for instance); learning disabilities (dyslexia or Attention Deficit Hyperactivity Disorder); and mental illness (depression, bipolar disorder etc.). Furthermore, as Davis (2005) points out, while all disabled people must actively seek the assistance they need, the invisibly disabled are often subject to greater scrutiny and the need to prove to others they have a legitimate disability. Such interrogation then may result in stress or failure to seek needed assistance as a means of bias avoidance (Olney and Brockelman, 2003).

Geography, disability and stigmatization

Park, Radford and Vickers (1998) chronicle geography's early engagement with disability. They identify three primary threads within the literature at the time: Geography and Physical Disability; Geographies of Sensory Impairments; and Geography and Mental Health. The first is largely focused upon as visible bodily impairments and how the physically disabled interact within their environments and the circumstances in which particular environments foster inclusiveness or exclusiveness. Geographic studies of sensory impairment occupy a somewhat liminal space between visible and invisible disabilities in that visual or hearing impairment may only become apparent to others with interaction or the use of visible assistive devices such as a hearing aid, white cane or leader animal. Early work in this field explored the spatial patterns of impairment often as a result of disease, while more contemporary efforts have explored such topics as how sensory impairment impacts mobility as well and how spatial cognition is affected by visual impairment (e.g. Golledge, 1993; 1997). Geographies of mental health have firmly delved into the field of invisible disability – one that society may not immediately discern or may be known only if the impairment is disclosed. Geographers have explored how the public disclosure of what may have been a private matter, namely mental illness, may result in social and spatial stigmatization within rural and urban environments (e.g. Dear, Taylor and Hall, 1980; Parr and Butler, 1999; Parr and Philo, 2003). In addition, as my experience suggests, there is a hierarchy of stigmatization that depends upon the particular mental health diagnosis.

Since those beginnings, the theoretical approaches to geographic research have broadened to include multiple definitions of disability, including viewing disability in terms of embodiment which integrates conceptualizations of disability as both biomedical and social in nature (Castrodale and Crooks, 2010). Disability is experienced socially, but also through the most intimate of spaces, the human body (Hansen and Philo, 2009). Disability as a social construction, Dear and colleagues (1997) note, is premised upon a person's condition, most importantly the sociospatial context in which it exists. Whether a person is regarded as disabled then, may be in part due to whether they are in a particular physical or social environment (see also Moss and Dyck, 2003). When a person and locational norms are in conflict, a boundary is established and that person becomes an Other. Chouinard (1997) calls for a critical evaluation of the ways in which places are defined by the able-bodied. She challenges the way in which 'lived environments incorporate and perpetuate physical and social barriers to the participation of disabled persons in everyday life', including the 'subtle and not-so-subtle reactions to disabled people that challenge their right to be and, in particular, to be in able-bodied, spaces' (Chouinard, 1997, p.380). This latter statement is particularly relevant in matters relating to mental illness within the university where mental capacity and intellect are deemed the crucial for success.

Within the context of higher education, learning disabilities among students have received significant attention. Efforts by Bolt and colleagues (2011) and Barnard-Brak and colleagues (2010), for example, highlight the role that negative perceptions held by both external actors (professors and other students) as well as the disabled students themselves, play in a university student's decision to disclose their learning disability and then both seek and use available classroom accommodations. Similarly, Luna (2009) explores the discourse surrounding the presence of learning disabled students at an Ivy League institution – a situation that many observers consider a contradiction in terms. While the university she describes claims an official policy of non-discrimination, formal and informal campus discourse with respect to learning disabilities reinforces the notion that students with them are intellectually deficient or are simply seeking a label and accompanying accommodation that would give them an unfair advantage in an academic environment.

The experiences and perceptions of university students with mental health conditions have been explored in both scholarly literature (e.g. Knis-Matthews, et al., 2007; Martin, 2010) and trade publications (Patton, 2012); however, scant attention has been directed to those issues as they relate to faculty with mental illness in particular. What does exist largely casts a wide net – addressing all forms of disability among faculty with a primary focus upon adherence to the federal Americans with Disabilities Act (ADA) by employers (e.g. Steinberg, et al., 2002; AAUP, 2012). There is, however, a small and growing literature exploring the personal experiences of disabled faculty themselves (e.g. Golledge, 1997; Tidwell, 2004;

Teghtsoonian and Moss, 2008; Moss, 2013) and that draws attention to the role that autobiographic approaches can contribute to more completely understanding mental illness in the academy. Autobiographic approaches seem particularly valuable in the academy where identifying faculty participants seems all but impossible due to stigmatization.

Autobiographic practice in disability studies

Parallels have been drawn between theorizations of disabled identities and queer theory. McRuer's (2003) work in many ways echoes that of Chouinard. In particular, he highlights the way in which not only the able-bodied but also heterosexuality are normalized as non-identities, those without some type of perceived deviance. Sherry (2004) and Samuels (2003) describe the similarities between the experiences of homosexual and disabled people (particularly invisible) and suggest that for both groups coming out can serve as an emancipatory process.

As relevant now as when originally declared by Parr and Butler (1999) over 15 years ago, is the need for an examination of illness, impairment and disability that includes lived experience. Autobiography then may serve as a form of empowerment for people with disabilities by allowing the disabled to claim their own voice in matters relating to their treatment (Atkinson and Walmsley, 1999; Garland-Thomson, 2005; Price, 2011). Valentine (2003) similarly points to the value of allowing disabled research participants to not only speak for themselves but also play a role in the dissemination of the findings. As a disabled researcher then, while I may potentially jeopardize my position within higher education by the public presentation of this project and my lived experience with mental health struggles, like Moss (1999) and others, I am gradually finding my voice and allowing the story of my invisible disability to be told by me rather than appropriated by others.

Experiencing invisible disability in the academy as place

My professional credentials were called into question due to my disclosure of my invisible disability a few years ago. In a closed-door faculty meeting, my colleagues and I were discussing the academic difficulties of one of our students. The student's advisor justified the academic struggle by noting it was predicated upon clinical depression. Like me, the student had never sought formal status by declaring a disability with the Office of Accessibility. I had actually known this student for quite a few years and during that time we had discussed how such a condition affects both our lives in an effort to reassure her that it was possible to achieve a balance between academic success and a positive mental health state. I relayed to my colleagues my long acquaintance with the student and suggested that placing additional academic burdens upon her would likely exacerbate the condition and not only would academic progress be interrupted, but the student's personal well-being would be jeopardized. After a tense discussion among all of us, my concerns were more or less dismissed and the meeting concluded.

While I continued to worry about the student, I thought little of the revelation of my mental health diagnoses with respect to the workplace. I had never asked for any kind of accommodation due in part to two things: I have been fairly vigilant in maintaining a regular relationship with a therapist and in taking antidepressant medication. I have had periods in my life where I felt that I could deviate from that pattern, but ultimately I have found that I am most content when both are in my life. Until that meeting, however, I had disclosed my condition only to my closest friends and family and to the student described. Even after my admission, I discussed it as little as possible at work. It was fine to tell my department chair or colleagues that I couldn't meet at a particular time because I had a physician's appointment, but I wasn't comfortable telling them I would arrive to campus a bit later to attend my biweekly therapy appointments. That might remind them that I was different, disabled and thus unable to cope emotionally or intellectually with the demands of my faculty position.

As the calendar year progressed, the overall climate within my department deteriorated. We had recently expanded our programme with an additional degree and disagreements arose about how that programme should develop. In particular, one colleague seemed to bully the rest and had formed a coalition with a staff member, fostering unrest and division among the faculty as well as our students. I had only been tenured that fall, so up until that point I had often looked the other way at unprofessional behaviour for fear of it having an impact on my tenure bid. One morning, the divisions in the department pushed me to enter into a verbal argument with the staff member who had been actively driving the division within the programme. It was hardly my proudest moment; however, at the same time it was not much different to the behaviours I had previously seen occasionally erupt between other faculty members in the programme since I had arrived – except this time it involved me. Shortly after the verbal confrontation took place, I made an attempt to speak to the other person to reconcile our differences but was turned away. I immediately notified my department chair of what had transpired.

I thought little of the matter as my daily interactions with that person returned to their usual uncomfortable but tolerable circumstances. Almost two months later I was informed by my department chair that the other person had written a complaint to the Office of Human Resources (HR) about my actions that day. I wasn't so much bothered by the act of complaining, but rather the content of her letter, which suggested my behaviour that day was likely tied to my publicly declared 'clinical depression' but could in fact be 'bipolar disorder' in the lay opinion of the writer. I was also informed that because the person who wrote the letter was interviewed by HR and declared that they felt threatened by me and feared physical harm, my name would be referred to the university Threat Assessment Team, which was in some way required to report such concerns to the U.S. Department of Homeland Security. I was astounded. Not only had my depression – which I had revealed in a closed meeting in the interest of helping a student – been transmuted into bipolar disorder, indicating the hierarchy of stigma with regard to diagnosis, my fitness for being on a university campus had been called into question.

Through the course of the investigation into the complaint, I was also made aware that the colleague who had allied with the staff member had been also trying to convince my department chair that I was bipolar – bringing him internet printouts highlighted with all the symptoms he felt characterized me. Furthermore, the faculty member suggested, as a bipolar person I was unstable and a threat – apparently even more so than had I simply been a depressed person. Not once did he ever attempt to discuss his concerns with me directly. Thankfully, my chair, who was aware of the hostile climate issues in our programme, did not take such claims as legitimate, but it did little to make me feel better as I was still left to deal with HR. My professional reputation based upon intellectual capabilities had been undermined. At the time of this event, my name was associated with over $8 million in federal research grant proposal submissions under review. With the mere mention of Homeland Security, I thought of the ways in which not only might I be professionally penalized, but also my collaborators. Was there a No Fund list for grant proposals like the federal government's anti-terrorism No Fly list? And if so, what would this mean for my career? How could I explain my situation to current or future research collaborators? Was my potential to receive federal research grant money – a scholarly expectation in my department and college – effectively blocked for the rest of my academic career?

I saw that my only recourse to address the accusations and rumors was to seek a remedy through the Americans with Disabilities Act compliance office on campus, which after some stalling and then another several months of investigation (that ensued only after I inquired whether I should redirect my complain to the Federal Equal Employment Opportunity Committee office instead) found that I had been harassed as a disabled person. Throughout the process, I feared that my disability would be further revealed and I would become known by my large numbers of undergraduate students as the *crazy professor*. Only a year before my incident, a shooting had taken place at a nearby university in which a faculty member who was found to be guilty as the perpetrator was defended on the basis of insanity. I was concerned that if my depression and eating disorder became widely known, I would be stereotyped into a category with her, and grow into a threat that would entail more scrutiny. I've always had a good classroom relationship with my students and colleagues elsewhere on campus and was afraid of those relationships being jeopardized.

As Dear and colleagues (1997) illustrate, research associated with the relative acceptability of disabilities indicates that mental illness is consistently popularly viewed as one of the least accepted disabilities across society at large – with even cancer and sensory impairments being regarded as preferable conditions. Disability researchers argue that when viewed solely from a biomedical framework, disabling conditions can be eliminated through prescribed medical care or intervention (Potts, 2008; Stone, 2008). Mental illness, chronic in nature, perhaps may then be viewed by the general public as so embedded within a person's mind that there is no chance for cure. Such hierarchies of disability and illness are interwoven within what Dear and his collaborators (1997) describe as the dimensions of acceptance. With respect to mental illness in the academy, two have particularly significant implications for my experiences: (1) unpredictability and dangerousness; and (2) functionality.

As Price (2011) suggests, my disability was unacceptable within the context of a university because my accusers managed to convince others that it might result in me behaving erratically or in ways that would physically harm others. But even more importantly, my mental illness itself was unpredictable and could have mutated into something my accusers deemed far more sinister. I was effectively declared emotionally, socially and ultimately intellectually unable to handle the professional career of a university faculty member. While some may suggest that my experience was premised upon the actions of two rogue employees, I would disagree. The university pursued the claims of my accusers and began penalizing me before even speaking to me; the penalties were implemented based on the notion of mental illness being incompatible with a university environment.

Six months after these events transpired, I used the experience as a means to begin a new path of research by presenting a paper at a regional geography conference that revealed my disability to the wider profession in an effort to move myself forward both professionally and personally – a site that Liggins, Kearns and Adams (2013) might describe in terms of Oldenburg's (1991) 'third place' where professional and personal relationships would be renewed or initiated. While so-called professional conferences might seem on the surface solely to serve career development, they also serve as a personal escape from my daily lived experience at home and at work. In contrast to focused conference sessions in which autobiography and the accompanying vulnerability are accepted, the small size of the conference required that I present in a conference committee-organized session attended not only by those who dropped into the session for other papers not directly related to invisible disability and the conscious inclusion of the self in research but also my own students, who came to support me and find out more about my new line of research. Despite having confidently submitted my abstract months earlier, my decision to reveal my personal self publicly again troubled me to such extent that I almost withdrew my paper. The unease I felt, however, perhaps highlights the need for research on the stigmatization of mental health illness among faculty, but also importantly that autobiography might be the only currently accessible means to conduct it.

Works cited

AAUP (American Association of University Professors), 2012. *Accommodating faculty members who have disabilities report*. Committee A on Academic Freedom and Tenure. [pdf] Available at: www.aaup.org/NR/rdonlyres/49CCE979-73DF-4AF4-96A2-10B2F111EFBA/0/Disabilities.pdf [Accessed 23 May 2012].

Atkinson, Dorothy and Walmsley, Jan, 1999. Using autobiographical approaches with people with learning difficulties. *Disability and Society*, 14(2), pp.203–16.

Barnard-Brak, Lucy, Sulak, Tracey, Tate, Allison and Lechtenberger, DeAnn, 2010. Measuring college students' attitudes toward requesting accommodations: a national multi-institutional study. *Assessment for Effective Intervention*, 35(3), pp.141–7.

Becker, Marion, Martin, Lee, Wajeer, Emad, Ward, John and Shern, David, 2002. Students with mental illnesses in a university setting: faculty and student attitudes, beliefs, knowledge, and experiences. *Psychiatric Rehabilitation Journal*, 25(1), pp.359–68.

Bolt, Sara E., Decker, Dawn M., Lloyd, Megan and Morlock, Larissa, 2011. Students' perceptions of accommodations in high school and college. *Career Development for Exceptional Individuals*, 34(3), pp.165–75.

Braddock, David L. and Parish, Susan L., 2001. An institutional history of disability. In: Gary L. Albrecht, Katherine D. Seelman and Michael Bury, eds., *Handbook of disability studies*. Thousand Oaks, CA: Sage. pp.1–68.

Castrodale, Mark and Crooks, Valorie A., 2010. The production of disability research in human geography: an introspective examination. *Disability and Society*, 25(1), pp.89–102.

Chouinard, Vera, 1997. Making space for disabling differences: challenging ableist geographies. *Environment and Planning D: Society and Space*, 15, pp.379–87.

Davis, N. Ann, 2005. Invisible disability. *Ethics*, 116(1), pp.153–213.

Dear, Michael, Taylor, S. Martin and Hall, G.B., 1980. External effects of mental health facilities. *Annals of the Association of American Geographers*, 70(3), pp.342–52.

Dear, Michael, Wilton, Robert, Lord Gaber, Sharon and Takahashi, Lois, 1997. Seeing people differently: the sociospatial construction of disability. *Environment and Planning D: Society and Space*, 15, pp.455–80.

Garland-Thomson, Rosemarie, 2005. Feminist disability studies. *Signs*, 30(2), pp.1557–87.

Gibbs, Nancy, 2011. The growing backlash against overparenting. *Time,* [online]. Available at: http://www.time.com/time/magazine/article/0,9171,1940697,00.html [Accessed 19 June 2012].

Golledge Reginald G., 1993. Geography and the disabled: a survey with special reference to vision impaired and blind populations. *Transactions of the Institute of British Geographers*, 18(1), pp.63–85.

Golledge, Reginald G., 1997. On reassembling one's life: overcoming disability in the academic environment. *Environment and Planning D: Society and Space*, 15, pp.391–409.

Hansen, Nancy and Philo, Chris, 2009. The normality of doing things differently: bodies, spaces, and disability geography. In: Tanya Titchkosky and Rod Michalko, eds., *Rethinking normalcy: a disability studies reader*. Toronto: Canadian Scholars' Press. pp.251–69.

Knis-Matthews, Laurie, Bokara, Josephine, DeMeo, Lorena, Lepore, Nichole and Mavus, Lauren, 2007. The meaning of higher education for people diagnosed with mental illness: four students share their experiences. *Psychiatric Rehabilitation Journal*, 31(2), pp.107–14.

Laurier, Eric and Parr, Hester, 2000. Emotions and interviewing in health and disability research. *Ethics, Place and Environment*, 3(1), pp.98–102.

Liggins, Jackie, Kearns, Robin A. and Adams, Peter J., 2013. Using autobiography to reclaim the 'place of healing' in mental health care. *Social Science & Medicine*, 91, pp.105–9.

Lum, Lydia, 2006. Handling 'helicopter parents.' *Diverse: Issues in Higher Education*, 23(20), pp.40–42.

Luna, Cathy, 2009. 'But how can those students make it here?': Examining the institutional discourse about what it means to be 'LD' at an Ivy League university. *International Journal of Inclusive Education*, 13(2), pp.157–78.

McRuer, Robert, 2003. As good as it gets: queer theory and critical disability. *GLQ: A Journal of Lesbian and Gay Studies*, 9(1–2), pp.79–105.

Martin, Jennifer Marie, 2010. Stigma and student mental health in higher education. *Higher Education Research and Development*, 29(3), pp.259–74.

Moss, Pamela, 1999. Autobiographical notes on chronic illness. In: Hester Parr and Ruth Butler, eds., *Mind and body spaces: new geographies of illness, impairment, and disability*. London: Routledge, pp.155–66.

Moss, Pamela, 2013. Becoming-undisciplined through my foray into disability studies. *Disability Studies Quarterly*, [e-journal] 33(2). Available at: http://dsq-sds.org/index [Accessed 11 November 2015].

Moss, Pamela and Dyck, Isabel, 2002. *Women, body, illness: space and identity in the everyday lives of women with chronic illness.* Lanham, MD: Rowman and Littlefield.

Moss, Pamela and Teghtsoonian, Katherine, eds., 2008. *Contesting illness: processes and practices.* Toronto: University of Toronto Press.

Oldenburg, Ray, 1991. *The great good place.* New York: Marlowe and Company.

Olney, Marjorie F. and Brockelman, Karin F., 2003. Out of the disability closet: strategic use of perception management by select university students with disabilities. *Disability and Society*, 18(1), pp.35–50.

Park, Deborah C., Radford, John P. and Vickers, Michael H., 1998. Disability studies in human geography. *Progress in Human Geography*, 22(2), pp.208–33.

Parr, Hester and Butler, Ruth, eds., 1999. *Mind and body spaces: geographies of illness, impairment, and disability,* London: Routledge.

Parr, Hester and Philo, Chris, 2003. Rural mental health geographies of caring. *Social and Cultural Geographies*, 4(4), pp.471–88.

Patton, Stacey, 2012. Colleges struggle to graduate students in distress. *Chronicle of Higher Education*, [online] 16 August. Available at: <http://chronicle.com/article/Colleges-Struggle-to-Respond/133699> [Accessed 18 August 2012].

Potts, Annie, 2008. The female sexual dysfunction debate: different 'problems,' new drugs – more pressures? In: Pamela Moss and Katherine Teghtsoonian, eds., *Contesting illness: processes and practices.* Toronto: University of Toronto Press. pp.259–80.

Price, Margaret, 2011. *Mad at school: rhetorics of mental disability and academic life.* Ann Arbor: University of Michigan Press.

Samuels, Ellen Jean, 2003. My body, my closet: invisible disability and the limits of discourse. *GLQ: A Journal of Lesbian and Gay Studies,* 9(1-2), pp.233–55.

Sherry, Mark, 2004. Overlaps and contradictions between queer theory and disability studies. *Disability and Society,* 19(7), pp.769–83.

Steinberg, Annie G., Lezzone, Lisa I., Conill, Alicia and Stineman, Margaret, 2002. Reasonable accomodations for medical faculty with disabilities. *JAMA*, 288(24), pp.3147–54.

Stone, Sharon Dale, 2008. Resisting an illness: disability, impairment, and illness. In: Pamela Moss and Katherine Teghtsoonian, eds., *Contesting illness: processes and practices.* Toronto: University of Toronto Press. pp.201–17.

Teghtsoonian, Katherine and Moss, Pamela, 2008. Signaling invisibility: risking careers? Caucusing as an SOS. In: Michelle Owen and Diane Driedger, eds., *Dissonant disabilities: women with chronic illness explore their lives.* Toronto: Women's Press. pp.199–207.

Tidwell, Romeria, 2004. The 'invisible' faculty member: the university professor with a hearing disability. *Higher Education*, 47, pp.197–210.

Valentine, Gill, 2003. Geography and ethics: in pursuit of social justice – ethics and emotions in geographies of health and disability. *Progress in Human Geography*, 27(3), pp.375–80.

10 Are we sitting comfortably?

Doing-writing to embody thinking-with

Kye Askins

Preface

This chapter explores the ways in which working with Qi (energy) and Traditional Chinese Medicine (TCM) theory is challenging and informing my academic praxis, through recognition of my own and other bodies and feelings as integral to processes of learning, teaching, research and writing. As an academic, I'm interested in understanding and explaining the role of emotions in social and spatial relations, but I've always struggled to think and write about emotion, affect and intimate geographies in academic forums.

My approach to emotions is through a feminist frame, conceptualizing mind and body as interconnected, and knowledge as – in complex, difficult and partial ways – embodied. This resonates with much of the significant effort being made to understand emotions and affect as non-tangible entities that are, arguably, both relational and embodied. These efforts range from written theoretical explorations, to empirical research which endeavours to engage with the emotional (for example see work in the journal *Emotion, Space and Society*). Indeed, attempts are being made across the social sciences to address difficult issues regarding how, as scholars, we may ask people to reveal and speak about that which is felt, how we might systematically understand this, and how we can re-present any analysis we make. Yet the predominance of writing, debating and professing emotions, affect and the intimate with (only) words seems and feels to me to be uncomfortable, frustrating . . . like a gap or an itch.

The core of my vexation is that normative academic ways of working and knowledge production already implicate me in trying to write that which is felt, in what remains a largely Enlightenment-inflected context of thinking and writing with (only) intellect (despite ever-growing critique). I stumble over words trying to make sense of sense, and my body seems sidelined in such efforts. So I decided to try to get at emotions and affect in the flesh, and started a three-year course in Zen shiatsu (hereafter shiatsu), explicitly to enrich my scholarly approach rather than as something outside or additional to it. Shiatsu, which translates as finger pressure, is a form of holistic bodywork in which (no more than) body weight is used in massage by a practitioner to support and strengthen a client or patient's own ability to balance themselves physically, emotionally and psychologically (see http://www.shiatsusociety.org/). It is a hands-on approach to well-being often described as a complementary therapy. More critically for me is that it draws on perspectives beyond Western epistemologies, that attend to flows of energy between body-and-mind, and between body-minds and the wider material world, through which I am trying to experientially learn how embodied knowledges may be grasped.

One of the things I'm realizing is that . . .

. . . My forehead is telling me something . . .

. . . by demanding my attention. For the past 18 months, most mornings I do a set of *Macca Ho* stretches and exercises (http://www.sohoshiatsu.com/Exercises. html), followed by five minutes lying on my back to check in with myself. This routine is central to the shiatsu course. As I'm lying down, the heels of my hands or fingertips often intuitively tap on a part of my body, and for the past few months they have been drawn to my forehead, tapping from just above between my eyes in a line up to the crown of my head, then back to the middle of my eyebrows, working outwards along the brows themselves. This tapping is central to another practice in shiatsu training, *Do-In* exercises (pronounced doe as in female deer; see https://www.youtube.com/watch?v=7FDUmbOiYP8). This practice stimulates the elemental meridians (air, earth, fire, water, wood) central to shiatsu and its TCM epistemological underpinnings. *Do-In*, like shiatsu itself, is about physically ena-bling the body to invigorate Qi (energy) movement through the meridians. In my experience, the act of tapping precipitates quite profound connections with my body, and leaves me somehow lighter, sometimes with a slight tingling sensation. TCM explains these feelings as the smooth flow of Qi (Beresford-Cooke, 2012).

The tapping also foments moments of insight; thoughts present themselves sud-denly and clearly, or slowly and requiring further consideration. In my forehead's case, I think the need to pay attention here has something to do with changing jobs six months ago. I moved from a teaching-intensive post, where I had been for nine years, to a research position at a different institution. For personal reasons, I'm not yet able to move cities, so my new job currently entails being away from home for several days and nights a week. Quite a change in lifestyle, routine, ways of working and engaging with colleagues, and, in particular, far more read-ing, thinking and writing time, all of which have attendant effects. I'm learning that connecting with Qi flow across my forehead is about a need to recalibrate, rebalance, find a new equilibrium, especially between body and brain, and enable movement between, across and through them.

My forehead is telling me something. I'm listening carefully, letting this knowl-edge surface, literally up through my skin *and also* in my mind. I'm trying to make sense – make meaning – from embodied senses. And all this is really difficult to express in (only) words. Thus, this chapter interweaves facilitation of an embodied exercise through its writing, *indicated in the following sections of italicized text*, in an attempt to foreground thinking and knowing as involving bodies as well as minds. You may choose to engage with this in your own way: doing-reading straight through the chapter, or picking out the exercise in its entirety at the start or end.

Sit in a relaxed and comfortable, yet upright, position. This may be on a chair or on the floor. If the latter ensure that your position may be held comfortable for a few minutes.

Gently rest your hands on your thighs, the chair seat or ground beside you, or the arms of the chair.

Begin to notice your breathing, as it currently is, for a few moments.

Now concentrate on how this breath comes in through your mouth/nose, and moves down into your chest. Let your mind follow the inhale and exhale back out into the air. Try to focus on your breathing, though letting any thoughts continue to run through your mind, not focussing directly on them or trying to block them out; rather paying attention to your body in breathing.

I need to inhale more deeply . . .

. . . came to me as a distinct realization while doing *Qigong* exercises, another aspect of shiatsu training (http://www.qigonginstitute.org/html/GettingStarted. php). I suddenly *knew* that I've spent much of my life breathing out, sighing or gasping for air. *Qigong* involves focusing on breath as linked with all parts of oneself and the world around you. On this particular occasion, I abruptly understood – mentally and physically – that I'm really good at exhaling, less so at inhaling. Also instantaneously, I grasped that this is embedded in gendered norms and practices: that my breathing out is connected to supporting, nurturing, giving to others before myself. Despite my struggles against sociocultural upbringing and dominant patriarchal structures, I still take on caring roles, and/or have such roles put upon me, in my private and professional lives.

This comprehension of breathing deeply, and what my forehead is telling me, are experiences of embodied learning, or *coming to know*. That is, the emergence of knowing through a connected mind-and-body, rather than knowledge production sought through thinking in (only) words, through theory, debating and professing. I use the term coming to know to emphasize that such learning continually *incorporates* my body into the frame, to hint towards a dimension of knowledge mostly lost or hidden or not-yet-found in Western normative structures of academia.

There are three aspects of an embodied coming to know that pose distinct questions to my academic praxis. First is precisely learning at and through the personal. In paying careful attention to my own body and how it feels in the world, I am stumbling upon or unearthing emotions, certain states of being, or links between feelings and events, that I had forgotten about or not-yet-realized. Beresford-Cooke (2012) argues that intimate memories are stored in bodies, in tissues-muscles-nerves, blocking Qi movement and affecting how one understands oneself and the world. Thus working with Qi in myself can enable the surfacing of hitherto unconsidered feelings and knowledges, and reshape my perspectives.

Interestingly, I experience a direct parallel in coming to know through Qi bodywork as I do in the process of writing. Setting out to script a text, I often think I know what I'm going to say, but it is in the writing itself that new perception and knowledge emerge. As DeSalvo (2000) points out, moments of profound insight can come from writing since they precipitate an awareness about ourselves, our relations to others and our place in the world that otherwise may lay dormant. Bringing bodywork and writing together could be a potent way to explore and come to new dimensions of knowledge. And it strikes me that writing, or more broadly communicating, with bodies-as-interconnected-with-intellect is long established in performance studies – what might this mean in social sciences? (How) Can I encapsulate embodied knowledges into writing? My attempts to do so through this chapter feel partial, limited, still frustrating . . . minor attempts to bridge the gap or scratch that itch.

Second, the use of touch in shiatsu is about relational energy connections with other bodies, and paying attention to others' physicalities. Specifically, shiatsu training centres on students listening to what happens to themselves when absorbed in Qi bodywork. From the first training weekend, students start giving shiatsu to volunteers (usually friends and family) to gain literally hands-on, embodied experience. This is exciting, nerve jangling and surprising. Students are also required to receive shiatsu from qualified practitioners regularly. As both giver and receiver, I've come to appreciate quite detailed things through touch, about myself and others: I've experienced energy connections and sensations that have (at various rates) percolated through and formed thoughts I can articulate. This learning with-and-about-others also unfolds, surfaces, in the same way as coming to know at the personal level; sometimes suddenly crystallizing in a specific sentence, sometimes as a more nebulous idea or image that has to be worked out in the speaking and explaining of it, to oneself or another.

In this second aspect, shiatsu for me corresponds to academic and feminist understandings of the self as relational. McNicholls (2014) conceives shiatsu sessions as meetings, in which connection through touch enables rebalancing of Qi in the receiver precisely because touch is an intimate association, an encounter, an exchange and a dialogue. Shiatsu does not diagnose in the English meaning of the word; someone's physical state is not to be corrected but enabled. Practitioners do this through shiatsu touch, practising unconditional, positive intersubjective relations. Crucially, the terms used regarding the therapeutic setting – giver and receiver – are recognized as problematic, since English cannot adequately encapsulate the interconnected nature of shiatsu. Even the translation *interconnected* is inadequate, since it suggests some separation in order to be then connected. TCM conceives body-and-mind, bodies-and-bodies, bodies-and-world holistically, *as Qi* (and I'm often reminded when reading about TCM how central language is in constructing our world views and possibilities, see Phipps and Kay 2014). Likewise, while Qi is often translated as vital energy, it is not force added to matter, but the state of being of any phenomenon. That is, in TCM there is no distinction between matter and energy: Qi is the 'thread connecting all being . . . the cause, process and outcome of all activity/growth/change . . . the propensity or inclination of all things' (Kaptchuk, 2000, p.8). Developed from early Daoist thought, Qi explains relationships and patterns of life as 'not only a set of correspondences, they also represent a way of thinking' (p.8).

Therefore training emphasizes listening to myself-in-connection-with-all in preparation for the *being-with* of shiatsu touch. I must try to understand or at least be aware of what I'm feeling-thinking in order to not project onto another-body's flow of Qi, or confuse what I may hear. I'm learning to learn holistically. And this is raising questions around body and energy awareness in academic contexts. What, if anything, should I incorporate in research and pedagogic approaches, discovered through (my) tissues-muscles-nerves? What are the implications for academic knowledges? I'm minded here of Bondi's (2014) treatise on the potential correlations between psychotherapy and research, and what a psychotherapeutic training and practice brings to processes of reflection and analysis (see also Boden, et al., 2016). I'm uncertain and uncomfortable about all this: boundaries blur beyond scholarly notions of positionality. How might I integrate embodied, Qi-focussed listening and coming to know through wider academic working practices?

In the research context, I'm not considering using shiatsu as a method (ethics committees may implode and the level of training and experience would be far beyond my capacities), yet the wiser shiatsu practices of intersubjective relations are opening up new ways of being with research participants, and alternative understandings of power relations and nuanced non-verbal communication. Probyn (2010) calls for research as being-with, in the space of the intimate, and I'm wondering about the implications when taken to embodied levels. Such issues are relevant to supervision and other meetings with colleagues and students too, as I'm coming to understand positions and conversations beyond intellectualizing. I am also more conscious of and can feel the ways in which I project energies-and-statements in lecture theatres, I'm more attuned to how I conduct and relate in seminars. This is not to say that I'm working more effectively or that these engagements are any improvement in how I work; such presumption would be anathema to TCM principles for a start. This is about a shifting awareness of my role in knowledge making and academic practices.

Third, as a geographer I'm conscious that academic spaces in which we discuss emotions, affect and energetic concepts (conferences, supervision, teaching) pay little heed, beyond theory, to our settings or what our bodies do there. Lea (2009) highlights how places and environments insinuate into wider ecologies of energy, how materiality and context are inherent in, for example, yoga practice. From a TCM perspective, given the importance of Qi flow and Qi as all things, shiatsu practitioners take care to enable harmonious (i.e., open to flow) energies in their treatment spaces. Likewise, the rooms used for the training course are places in which I feel – to the core of myself – supported and nurtured. The Qi of place, then, is inherent in coming to know, crucially together with our activities there.

Recognizing the Qi of place queries how space can be made, in which students, colleagues and I can attend to our bodies-in-relation in academic practice. I cannot redesign and refit university spaces, but I can reorder within them, rearranging and introducing things to disrupt scholarly work as (only) words. I can also focus on the activities themselves, and I'm tentatively introducing some *Do-In* and breathing exercises in workshops, conference sessions and relevant teaching, as a shiatsu-inflected nudge towards listening to more than intellectual thought, through which we may come to know (each other and the world) in more embodied ways. Such practice must carefully consider difference and power relations, and precisely their *bodied* dimensions, as part of addressing the dominant disconnect between bodies and intellect that enables all manner of -isms to be reiterated, rendered invisible through academic preponderance on minds alone (see Domosh, 2015) on the visceral knowing-ness of hurt. Any such activity should be designed to be negotiated with contours of diverse bodies-abilities – and the embodied exercise through this chapter is intended as adaptable.

Now move your attention to your feet. Are they comfortable? Are they relaxed? Do you feel any connection through them with the ground? Wiggle your toes slowly and move your feet if you feel you want or need to. Imagine your breath connecting from your chest down through your core, your thighs, through your knees, shins and calves, to your feet, and through them into the ground. Try to listen to how they are.

Stop and smell the roses . . .

. . . presented itself as a phrase in my thoughts after a training session towards the end of the first year of the course. Walking home at the time, I physically stopped on the street for several seconds, the idea and sentiment was so strong. I considered how I rush around ever faster-faster-faster; how work is go-go-go and home life a spinning-never-sit-still existence. I remembered my own academic call for 'slower, more engaged geographies' (Fuller and Askins 2007, p.599). Those thoughts resurface here, now, again, as I'm sitting uncomfortably at my computer, due to niggling tensions between completing this chapter, administrative and marking deadlines, email backlog, family and personal commitments, and wanting to find time to read a paper for slow scholarship (Mountz, et al., 2015). I'm also uneasy with myself as I skipped *Macca Ho* this morning to get started on all these tasks. Readers may know how this feels, and the myriad critiques regarding pressures of the neoliberal academy, and role of late capitalism in increasing health concerns across wider society (Schulte, 2014).

Taking time to breathe in . . . slow down . . . contemplate . . . connects with notions around stillness. Conradson (2010, p.73) finds that stillness is sought as counterpoint to the lived intensities of Western societies; as a recalibration attained through 'particular forms of individual and collective practice', including meditation and mindfulness. In such techniques, the focus is on being fully present, in the moment, able to direct attention away from the out there world to the in here body, and Conradson is clear that subjective experience of stillness can be through physical activity and movement as much as stasis (for example running or T'ai Chi). Certainly, many people choose shiatsu therapy as an antidote to or periodic escape from hectic lives.

I experience some semblance of this stillness through working with Qi in *Macca Ho, Do-In, Qigong* and shiatsu, and one of the most vital things I'm learning, theoretically and physically, is that stillness is not in binary opposition to (neoliberal) rushing-disjointedness. TCM conceives such states as *inherently twined* all the way through. Stillness is part of rushing which incorporates stillness. Embodied-mindful presence and the absence thereof are at the same time. In-the-moment and not here-and-now exist simultaneously. This perspective is enshrined in the central concept of Kyo-and-Jitsu: Kyo (that which is less energized) and Jitsu (that which is stressed or over-energized) as inherently coupled, relating through each other. Palmer (2013) warns of the risk of a looser translation, which describes Kyo as empty (a part of the body where practitioners sense a lack, or nothing) and Jitsu as full. Rather the original Chinese (Kanji) characters represent Kyo as a powerful energy that is hidden, and Jitsu as a symbol that means 'the same on the surface as underneath' (p.6). Thus Jitsu is a resource already owned by receivers, which they use to rebalance areas of Kyo, to become aware of and integrate unconsidered feelings-thoughts, while practitioners support this through giving shiatsu and connecting Qi.

There are again parallels with psychotherapy here, and this approach also resonates with academic (especially feminist and postcolonial) concepts around agency. Moreover, for me, learning-working with Qi enables – requires – an orientation towards *attention and intention* to people and things in *respect of* their Qi flow, which infers a certain politics. Slowing down . . . smelling roses . . . listening to muscles-tissues-nerves . . . are political acts; outwith the individualism of neoliberalism, inwith collectivism and relational care. This politics is not about taking a break from neoliberal lived intensities for a while, to rebalance, gain moments of stillness before diving back in. The purpose is to live in respect of Qi and all things, *to live differently*. This applies to yoga, meditation and other related philosophies too, though there is a spectrum of practice that includes the take-a-break-dive-back-in approach, as Conradson's (2010) research shows. I'm interested in moving beyond stillness or slowing down as a way to survive neoliberal pressures, and towards a more respectful, collective orientation – *living academia differently* . . .

Now move your attention to your hands. Are they comfortable? Are they relaxed? Wiggle your fingers slowly and adjust your hand positions if you feel you want or need to. Imagine your breath connecting through from your chest along your shoulders, down your upper arms, through your elbows, along your forearms and into your hands and fingers. Is there anything to notice that they may be telling you?

I am losing my sisters . . .

. . . is a nagging feeling that I've had for a few months now. I tried several times to voice this feeling to friends and family, to discuss and make sense of it, but it remained fuzzy. Yet in shiatsu it became clear – in receiving shiatsu recently, this interjection formed itself into a distinctive thought. Typing that thought into a sentence brought inrushes of emotion, embodied memories and awareness of profound shifts in relations with both my sisters. I've been losing them for several years, I realize. Not in the (poststructural) sense that we are all emerging, shifting, fluid subjectivities, becoming across time and space, thus always losing and gaining selves and others. This losing is specifically and significantly ruptured, knitted through separate and serious mental health issues; my sisters are markedly not who they were. When we meet, our relationships have become quite different, and there is no tidy story (Tamas, 2009) to tell about this.

Rather, my aim here is to reflect honestly, uncomfortably, on how the intimate is caught up within wider social discourses and spatial scales. Intimate realizations and moments trace themselves through writing, research, learning, and personal and professional relationships; my body is telling me not to bury them, not to stuff them back down into muscles-tissues-nerves. They are implicitly part of how I come to know. Thus, scripting my sisters here is a way to ghost what I'm encountering in current fieldwork, in which mental health issues comprise, in varying degrees, participant positions, experiences and identities. I am listening to empirical material in new and recalibrated ways, inherently related to how I'm making sense of losing my sisters. I need to reflect on this shift, appreciate its meaning in analysis, and what this kind of coming to know may mean to research questions and wider scholarly debates. Shaw (2013) discusses such intimacy as process, arguing for more conscious engagement with the ways in which researcher memories filter into research through auto-ethnography and autobiography, and other authors in this volume speak explicitly to such process.

Likewise, I am concerned with how relations with family and friends stretch across time and space, and become enmeshed in academic knowledges in myriad ways. Pain and Staeheli (2014) suggest intimacy as three coinciding sets of relations: spatial relations that extend from proximate to distant; modes of interaction also expanding personal to global-and-distant; and sets of practices connecting the body and (other bodies) far away. Critically, any hierarchical ordering of one spatial scale over another is destabilized. National and global policies and events are already caught up within the everyday and ordinary, such that as scholars, we should acknowledge the intimate and emotional in research and academic endeavour, to more fully understand social and spatial processes and relations. My point here is around how embodied engagements are as critical as conscious reflection – moreover how the embodied and conscious are intertwined in this.

Hence I am heedful of intimacy, of scripting my sisters as coinciding sets of relations and incorporating this in empirical research-and-analysis. There are also related, interscaled/ing issues regarding representing the intimate and embodied in writing, within the strict confines or norms of scholarly journals and texts. I feel vulnerable writing about Qi and shiatsu as academic praxis, because I have no evidence, yet I know there's something in this touch, finger pressure, holistic bodywork. I'm exceeding dominant structures of education, employment and wider socio-cultural milieu that I'm familiar with, and I'm tentative, stumbling, unsure of myself. Yet – and this is the point – writing is intimate and embodied. Why else would we feel nervous when submitting a paper, or gutted at rejection, or disappointed or even a bit sick at harsh critique? Our bodies are at the heart of this process.

Further, Kennedy (2013) considers how writing involves a range of necessary intimacies, as meanings, resonances, memories and emotions are fired in readers' minds, connecting across place and people. Not least, this occurs as we create relationships with texts (Ahmed, et al., 2000), as writers and readers. You are now beginning to weave yourself through this text, as I am woven in deciding what to say. Similarly, Bennett (2000, p.120) explains 'it is not just about you reading me, but being aware of yourself reading, feeling, and being touched by that feeling . . . It is not easily told, but far easier felt.'

Conceiving and approaching writing as surpassing words, as words-with-bodies, opens up those wider dimensions of coming to know I am searching for, attending to distanciated intersubjective relations, and aligning with the more respectful and collective orientation to academic praxis mentioned above. Indeed, Moss (2014) emphasizes the practice of writing as an affirmative politics, valuing potential and possibilities, working towards sustainable well-being in academic endeavour that embraces embodied selves in intellectual and social practices as a collective project. Shiatsu underscores such woven-ness, through *attention and intention*, through being-with and Qi-in-all-things. For me, this spurs critical reflection around how, as scholars, we review, critique and offer comment and feedback on peers' and students' work – with our *selves* (more than intellect), framing more holistic bodies of literature and work.

Through this chapter, perhaps we can consider breathing together, as together we are related through these words, what we think and how we feel about them:

Try to focus on your breathing, though letting any thoughts continue to run through your mind, not focussing directly on them or trying to block them out; rather paying most attention to your breathing.

Now move your attention to your head. Is it comfortable? Is it relaxed? Gently rotate your head, stretch your neck side to side, tip forwards and backwards if you feel you want or need to. Imagine your breath connecting from your chest up through your neck, around your face, skull, forehead and to the uppermost point of your head. And from there, connecting with the sky, letting thoughts emerge as they do.

I am air, earth, fire, water, wood . . .

. . . simultaneously and in ever-shifting ways, relations, im/balances. These are TCM elemental associations that run through our bodies' meridians (Beresford-Cooke, 2012). I am coming to know (feel-and-think) myself from this perspective. To comprehend myself in this manner feels intimate *and also* more immense, expansive, connecting to wider Qi and its essences – air, earth, fire, water, wood. Such perception sits alongside being-becoming woman, mother, partner, white, middle-aged, heterosexual and other such categories given or available to me in normative Western society and academia. I'm wary of fetishizing Zen shiatsu as an Other to be known, tamed, controlled and colonized, or setting diverse eastern philosophies uncritically together in some box. I am aware that elemental nomenclatures hold no more truth than those I've used as shorthand all my life. And I'm acutely aware that I'm making a leap of *faith* in shiatsu training, which is fascinating since I have never understood – or felt – social constructionism, or epistemology, in quite this way before. This is discomforting and destabilizing, urging and animating . . .

. . . interrupting usual, habitual, embodied patterns and altering entrenched horizons, socio-cultural heritages and positions. I'm attempting to access different dimensions of being-becoming, across my personal and professional lives. These efforts at doing-writing perhaps point towards the new energy geographies outlined by Philo, Cadman and Lea (2015), though vitally also the need to transcend theoretically-inflected accounts of affective energies or forces, to foreground senses as inextricably embodied in holistic body-minds. As an academic interested in understanding and explaining the role of emotions in social and spatial relations, I'm trying to incorporate physical sensations, feelings and thought, privileging none; to use energetic information to hold and engage a Qi field (Beresford-Cooke, 2012). There will always remain gaps in translation, socio-cultural norms regarding emotions, body language and so on, yet awareness is a first step to working through such gaps, of coming to know inclusive of bodies and feeling as well as social construction and theory . . .

. . . to nurture, and work-think-write in more holistic and sustainable ways. Working with Qi and TCM theory implicitly-and-explicitly entails constant processes of recalibration towards living collectively. TCM insists we concurrently nourish, replenish and respect ourselves alongside caring-for-about-with others: to open out and give, one needs to take in and receive. My underlying aim is to better balance breathing in and out, while:

Still breathing as you are, pay attention to your core. Try to feel your breath circulating from your chest, through your heart, down to your stomach, then abdomen. Breathing into the centre of yourself . . .

. . . letting any thoughts continue to run through your body-and-mind, not focussing directly on them or trying to block them out. Rather, noticing them as they surface, listening to what you are coming to know.

Works cited

Ahmed, Sara, 2000. *Strange encounters: embodied others in post-coloniality.* London: Routledge.

Bennett, Katy, 2000. Inter/viewing and inter/subjectivities: powerful performances. In: Annie Hughes, Suzanne Seymour and Carol Morris, eds., *Ethnography and rural research*, Cheltenham: Countryside Community Press, pp.120–35.

Beresford-Cook, Carole, 2012. *Shiatsu theory and practice.* London: Elsevier.

Boden, Zoë, Gibson, Susanne, Owen, Gareth and Benson, Outi, 2016. Feelings and intersubjectivity in qualitative suicide research. *Qualitative Health Research*, 26(8), pp.1078–90.

Bondi, Liz, 2014. Understanding feelings: Engaging with unconscious communication and embodied knowledge. *Emotion, Space and Society*, 10, pp.44–54.

Conradson, David, 2010. The orchestration of feeling: stillness, spirituality and places of retreat. In: David Bissell and Gillian Fuller, eds., *Stillness in a mobile world*. London: Routledge, pp.71–86.

DeSalvo, Louise, 2000. *Writing as a way of healing: how telling our stories transforms our lives*. Boston, MA: Beacon Press.

Domosh, Mona, 2015. How we hurt each other every day, and what we might do about it, [online] *American Association of Geographers' Newsletter*, 2015 (5 May). Available at: http://news.aag.org/2015/05/how-we-hurt-each-other-every-day/ [Accessed 19 April 2015].

Fuller, Duncan and Askins, Kye, 2007. The discomforting rise of 'public geographies': a 'public' conversation. *Antipode*, 39(4), pp.579–601.

Kaptchuk, Ted, 2000. *Chinese medicine: the web that has no weaver*. London: Random House Publishing.

Kennedy, Alison Louise, 2013. *On writing*. London: Johnathan Cape.

Lea, Jennifer, 2009. Liberation or limitation? Understanding Iyengar Yoga as a practice of the self. *Body and Society*, 15, pp.71–92.

McNicholls, Paul, 2014. Treating and meeting. *Shiatsu Society Journal*, 129, pp.28–31.

Moss, Pamela, 2014. Some rhizomatic recollections of a feminist geographer: working toward an affirmative politics. *Gender, Place and Culture*, 21(7), pp.803–12.

Mountz, Alison, Bonds, Anne, Mansfield, Becky, Loyd, Jenna, Hyndman, Jennifer, Walton-Roberts, Margaret, Basu, Ranu, Whitson, Risa, Hawkins, Roberta, Hamilton, Trina and Curran, Winnifred, 2015. For slow scholarship: a feminist politics of resistance through collective action in the neoliberal university. *ACME: An International e-Journal for Critical Geographies*, [e-journal] 14(4), pp.1235–59. Available at: http://acme-journal.org/index.php/acme/article/view/1058 [Accessed 19 April 2015].

Pain, Rachel and Staeheli, Lynn, 2014. Introduction: intimacy-geopolitics and violence. *Area*, 46, pp.344–7.

Palmer, Bill, 2013. The tiger in the grove: Kyo and Jitsu in Movement Shiatsu. *Shiatsu Society Journal*, 128, pp.5–9.

Philo, Chris, Cadman, Louisa and Lea, Jennifer, 2015. New energy geographies: a case study of yoga, meditation and healthfulness. *Journal of Medical Humanities*, 36, pp.35–46.

Phipps, Alison and Kay, Rebecca, 2014. Languages in migratory settings: place, politics and aesthetics. *Language and Intercultural Communication*, 14(3), pp.273–86.

Probyn, Elspeth, 2010. Introduction: researching intimate spaces. *Emotion, Space and Society*, 3, pp.1–3.

Schulte, Brigid, 2014. *Overwhelmed: work, love and play when no one has the time*. New York: Sarah Crichton Books.

Shaw, Wendy S., 2013. Auto-ethnography and autobiography in geographical research. *Geoforum*, 46, pp.1–4.

Tamas, Sophie, 2009. Writing and righting trauma: troubling the autoethnographic voice. *Forum: Qualitative Social Research*, [e-journal] 10(1), Article 22. Available at: http://www.qualitative-research.net/index.php/fqs/article/view/1211/2641 [Accessed 23 March 2015].

Part III

Multiple aspects of researching intimacy

11 Accelerating intimacy?

Digital health and humanistic discourse

Courtney Donovan

In recent years, the digital health sector has experienced explosive growth. The excitement over this area of innovation is evident in numerous arenas, including *The New York Times*, which ran a special issue on 'The Digital Doctor' in October 2012 (Anonymous, 2012). The articles in the special issue outline a range of issues: using apps for medical diagnosis, managing digital records and planning medical offices through the use of computer aided-design. One article suggests the possibilities for doctors to use GPS tracking to more intimately monitor a patient's behaviour. This last article goes on to describe that GPS offers a promising tool for closely following patients with serious chronic health conditions or who may be experiencing a health crisis.

Just a few years later, the digital health sector made what I call an empathetic turn. Terms like engagement, listening, empathy, experience and intimacy are now taking centre-stage in the media as well as in promotional materials for digital health startups (Wanshel, 2016). By using terms like these that convey emotion in the context of technologically-based care, it seems fair to say that Silicon Valley and its counterparts have sought to reorient the focus of industry away from functionality and efficacy to humanistic medicine, an approach to medicine that emphasizes the importance of demonstrating compassion and empathy in the provision of care. In this particular rendition of humanistic medicine, digital health suggests digital technologies can be used as a tool to enable providers to connect more deeply with patients beyond the medical office, improving the quality of patient experiences.

Indeed, there are a number of organizations focused on ways to use technology to enhance patient experiences. One example that stands out comes from Embodied Labs, an immersive virtual reality company. In an online video Embodied Labs shows young doctors donning a virtual reality headset to essentially become a patient named Alfred, in what is called an 'empathy training exercise' (Wanshel, 2016). Upon looking through the headset the provider suddenly experiences first-hand what it is like to be Alfred, a fictionalized older person. The program, We are Alfred, allows a doctor to suddenly situate him or herself directly in the life of an older patient. Through Alfred's eyes, the provider is pressed to understand immediately what it is like to try to interact at a family gathering or in a doctor's office while struggling with a number of ailments, including greatly diminished field of vision and dementia. In the same video demonstrating We Are Alfred, a doctor wearing a headset remarks that as Alfred, she can't concentrate and remember what a fictionalized doctor asks her, because she is too distracted. According to Embodied Labs, the ultimate goal of We Are Alfred is to foster empathy between care providers and their older patients.

Interest in developing approaches that humanize the medical encounter in part stems from the concern that technology and its proponents have focused largely on health metrics and measurement in advocating digital health to health providers. In particular, critics of digital health express apprehension over whether digital health offers doctors a quantitative understanding of patient health to the exclusion of other factors, including patient experiences or the psychological factors that help to inform health choices and outcomes (Webster, 2007). In this respect, the focus on humanistic medicine in digital health brings to mind an earlier critique of Western medicine led by feminist and other critical scholars of health and medicine (Clarke and Olesen, 1999). That is, the digital health sector's fascination with humanistic medicine echoes critiques that Western medicine overemphasizes the biomedical model and ignores experience and affect in the treatment of patients (Stingl, 2014).

In this chapter, I offer a preliminary treatment of the ways in which digital health engages with humanistic medicine. In particular, I trouble the ways in which the tech industry frames humanistic concepts as I briefly review how tech companies adopt the discourse of humanistic medicine. I first consider how humanism became a concern within medical and health scholarship. I next address strategies that medicine has adopted to address the perceived distancing of providers from patients. I then consider how humanistic thinking has recently been integrated into digital health. I position the perceptions of digital health companies alongside apprehensions expressed by critical digital health scholars. I conclude with thoughts and reflections for ways in which humanistic discourse and approaches might be reflexively integrated into digital health in the interest of listening and responding to health experiences, as opposed to pathologizing and marginalizing patients.

Humanism in medicine

In recent decades, humanistic medicine has become a pervasive term used in medical literature (Rabow, et al., 2016). The main premise underlying humanistic medicine is to address patients' feelings of dehumanization in clinical encounters. Much has been written about the sense of alienation and isolation patients feel in the context of the increasing demands on providers by healthcare systems (Charon, 2006). Patients feel like objects in a service assembly line, excluded from their own healthcare and separated from their own bodies. Western medicine, reliant on biomedicine as its knowledge, tends to focus on a limited range of medical issues and emphasize individualized body parts instead of looking at the body as a whole. Feminist scholars have been at the forefront of these critiques (Bordo, 2004; Mamo, et al., 2010). They contend that this particular construction of health and medicine has problematically focused on a limited range of health and medical issues. Biomedical framings of illness and disease have resulted in rather narrow conceptualizations of health that do not give adequate attention to the social context within which disease is identified and the multiple factors that may help to explain individual health experiences (Zeeman, Aranda and Grant, 2014). Thinking of the example of Embodied Labs' Alfred, there is concern that providers downplay the significance of emotional aspects of health and health care, including potential embarrassment over impaired hearing or diminished vision that may prevent a patient from asking for the care they need (Gordon, 1988; Gray, 2012).

The idea that patient experience should be given greater attention in personal encounters with healthcare professionals has slowly expanded beyond the realm of feminist critiques, into places such as medical schools and medical journals (Gray, 2012; Plant, et al., 2015). Patients' needs are being eclipsed by healthcare systems based on biomedicine. A steady stream of medical research since the turn of the twenty-first century has detailed the consequences of developing clinical practices based on the premises underlying biomedicine. Clinical practice, itself undergoing a transformation as a result of the intensification of record keeping, extensiveness of insurance billing and specialization of medical knowledge, considerably constrains interaction between a patient and a healthcare provider. While Western medicine has expanded its tools for evaluating and tracking patient health outcomes, such tools have played a role in distancing medical providers from patients. The reorientation of medicine towards technology and research perpetuates an environment in which caring and compassion become secondary goals (Shapiro, 2008). As part of a response to such critique, medical schools have turned their attention to developing and promoting relationship-centred or patient-centred care (Fiscella, et al., 2004). It is this emphasis on the patient, patient experiences and the emotional aspects of care that has spurred the interest in humanistic medicine in the hope of providing more compassionate care. Although there is little consensus about what exactly humanistic medicine is or how humanism can inform medicine, there are some common core characteristics that define such an approach, including for example an emphasis on providing compassionate and empathetic care (Cohen and Sherif, 2014) and the goal of making providers sensitive and responsive to the health needs and of patients.

One practice garnering widespread attention is narrative medicine. Narrative medicine underscores a more enhanced experience of illness and disease that emphasizes the ways people convey experiences of illness, disease and health, an approach that emphasizes the importance of different stories in conveying health and medical experiences. Narrative medicine draws from the literary techniques of literature, poetry, plays and other forms of textual material. As Charon (2006) explains, narrative medicine initially emerged as a field in response to some health practitioners recognizing their limits in understanding patient concerns. Practitioners acknowledged that their medical training had not prepared them adequately for the skills needed to understand patient struggles in, for example, chronic illness, aging and even death. They began seeking advice from those who understood the complexity of the experience of illness, including teachers, writers and patients, who also understood the mechanics of storytelling as a narrative practice. Narrative medicine has since grown to incorporate other techniques that convey the intricacies of the experience of illness in various forms.

Humanism and digital medicine

Just as the medical field is turning to the idea of humanistic medicine, companies working in the area of digital health have similarly oriented their attention towards patient-centred strategies (Edwards, 2016; Krist and Woolf, 2011). Digital health is an umbrella term for the use of diverse digital technologies in health care, which includes medical devices and information systems in doctor's offices and hospitals to track and manage patient health information. Digital health also encompasses actions that patients can take including using genomics technology, referring to personal fitness apps and relying on social networking and mobile connectivity, with the aim of interacting with their health provider. Digital health also refers to tools that can be used in medical education, including virtual reality tools, which situate students directly in the experience of potential patients who may be different from them. Digital health technologies therefore offer an opportunity to augment the provider's assessment of a patient's health and extend the medical gaze beyond the context of the clinic. In the context of digital health, patient-centred strategies are those that technologies and strategies that promote a partnership between providers and patients, in order to engage patients in their health maintenance.

The interest of digital health in humanistic medicine comes at a time when providers face increased scrutiny over patient-centred care under the Affordable Care Act (ACA) of 2010 in the United States. This manifests in two main areas. First, under the ACA, hospitals face fines if readmission rates on certain treatments exceed the national average (Orszag and Emanuel, 2010; Kocher and Adashi, 2011). Second, under the auspices of the ACA hospitals are rewarded for patient satisfaction. With so much at stake financially, it is little wonder that providers are looking for alternative ways to engage with patients.

Tech companies have responded to these ACA mandates by suggesting to providers that digital technologies can be employed to foster effective communication with patients. Effectiveness includes empathetic care (Klein, Hotstetter and McCarthy, 2014). Empathy figures prominently in the discourse of digital health and can best be summarized as the ability of providers to comprehend the meaning behind patient communication. The thinking goes that if patients are offered more opportunities to connect with their provider and to be heard, patient satisfaction and adherence to doctors' recommendations will increase.

Tech companies envision digital communications as an effective way to connect providers with patients, and also as an accessible method to engage patients beyond the clinical encounter. However, tech companies promote digital health as more than a bridging method in patient–provider relationships. Indeed, much of the current work in digital health centres on how providers can rely on technology as a method to make the clinical encounter more meaningful (Topol, 2012). Hence, tech companies promote digital health as a tool that allows the doctor to better understand and appreciate patient experiences. The use of instant messaging technology, remote monitoring devices and augmented virtual reality headsets, are just some of the many ways that the tech industry has called upon doctors to rethink the ways in which they can promote empathetic engagement with patients.

Tech company reports and promotional materials on digital health and empathy suggest the potential benefits of digital health in responding to patient health needs (Detmer, et al., 2008). Proponents of digital health argue that the use of digital platforms for patient-centred care provides the opportunity to collect a more comprehensive picture of patient health (Topol, 2012). The perception is that digital health offers more than a single snapshot of a patient's health status. Yet there is still concern about whether digital health could potentially replicate the same problems that catalysed the development of humanistic medicine in the first place. Although digital health strategies are intended to improve interaction between patients and providers, virtual interactions and relationships may instead foster impersonal interactions and distance between patient and practitioner (Shaywitz, 2012). Moreover, providers and patients alike have expressed apprehension that digital approaches will get in the way of compassionate interactions that emerge from human contact. Thus some hybrid of both may be of interest.

Critical reflections on digital health

The ascendance of digital health comes at the moment in which social interactions are often mediated through the filter of technology. It is not uncommon for individuals to connect online and carry on interactions that require minimal or nonexistent physical contact (Orton-Johnson and Prior, 2013). Many scholars of social media have taken on the task of exploring this paradox of what has been referred to as the intimacy of distance (Dill-Shackleford, 2011). Here, scholars refer to how individuals hold the idea of another in their minds, as they interact via digital devices (Gardner and Davis, 2013; Orton-Johnson and Prior, 2013). In effect, digital technologies rely in part on constructing the illusion of intimacy. Intimacy refers here to social relationships that provide the impression of being physically or emotionally close and pivots on an arrangement of proximity and vulnerability. While clinical relationships in healthcare are not always associated with emotional closeness, biomedical diagnosis has relied on drawing insights from physical nearness to the body (Foucault, 1973). Herein lies the problem. Because a health practitioner is physically close does not imply that they are actually hearing or listening. Indeed, biomedicine has relied on a particular ontology in which closeness does not require providers respond to the vulnerability experienced by a patient.

Humanistic medicine has attempted to interrupt a model of medical practice steeped in this idea of objectivity and distance standing in for proximate knowledge of the body. While the changes borne out of humanistic discourse are still unfolding, tech companies are positioning electronic empathy as an improvement to the practice of biomedicine. Indeed, at a couple of recent Meetups in San Francisco focusing on integrating humanism into digital health, speakers specifically addressed how digital health applications offered an opportunity to *accelerate* empathetic and intimate engagements between patients and providers (Hwang, 2016; Shlain, 2016). Acceleration is desired for it affords providers with a preconfigured connection that can more quickly establish rapport with their patients and thus get closer to them.

Although digital health offers the promise of improved patient-provider engagement, critical digital health scholars suggest the need to reflect critically and constructively on these strategies. While digital health companies may suggest otherwise, these scholars emphasize the importance of acknowledging that digital health and its adoption is not value-free or neutral (Lupton, 2015). Of particular interest to critical scholarship are the implications digital health technologies pose for social groups and individuals who have historically been marginalized, including those who are socioeconomically disadvantaged, older persons or individuals living with disabilities (Yamin, et al., 2011). It is already widely established that reliance on digital platforms, including the internet and downloadable apps, is strongly correlated with a number of demographic features including age, income, education level and geographic region. Consequently, certain individuals and groups of individuals may have limited access to digital technologies. Complicating access, there are some people that do not have the required health knowledge or digital literacy to make use of digital health tools (Neter and Brainin, 2012).

Digital health tracking also presents another iteration of neoliberal health promotion strategies (Johnson, 2014; Lupton, 2015). As a new cost-cutting opportunity digital health positions individuals to take responsible for their own health. But the interactive platforms do not permit ownership of health information about individuals for individuals; practitioners retain records to which individuals have limited access. Thus, digital health bears new questions concerning the role of health surveillance in personal, professional and system transparency, accountability and responsibility (Lupton, 2012). Much like earlier writings on personal responsibility and health critical digital health scholars identify the moralistic discourse used to describe individuals who opt not to rely on or simply do not know how to use digital health technologies (Ahlin, 2013).

Concluding remarks

Reliance on humanistic medicine forces digital health advocates to move from initial assumptions that medicine and technology are neutral fields, and instead recognize the ways in which intimacy and vulnerability become meaningful in relation to medicine and social media (Rice and Katz, 2000; Anderson and Agarwal, 2011). The use of humanistic discourse in the context of digital health rests on the supposition that patients will be more likely to engage with providers and be more actively involved in their health maintenance if they receive personalized texts inquiring about their health status. Digital health also recognizes the

Neter, Efrat and Brainin, Esther, 2012. eHealth literacy: extending the digital divide to the realm of health information. *Journal of Medical Internet Research*, [e-journal] 14(1), pp.e19. doi:10.2196/jmir.1619

Orszag, Peter and Emanuel, Ezekiel, 2010. Health care reform and cost control. *New England Journal of Medicine*, 363(7), pp.601–3.

Orton-Johnson, Kate and Prior, Nick, eds., 2013. *Digital sociology: critical perspectives.* New York: Palgrave Macmillan.

Plant, Jennifer, Barone, Michael, Serwint, Janet and Butani, Lavjay, 2015. Taking humanism back to the bedside. *Pediatrics*, 136(5), pp.828–30.

Rabow, Michael, Lapedis, Marissa, Feingold, Anat, Thomas, Mark and Remen, Rachel, 2016. Insisting on the healer's art: the implications of required participation in a medical school course on values and humanism. *Teaching and Learning in Medicine*, 28(1), pp.61–71.

Rice, Ronald and Katz, James, 2000. *The internet and health communication: experiences and expectations.* Thousand Oaks, CA: Sage Publications.

Shapiro, Johanna, 2008. Walking a mile in their patients' shoes: empathy and othering in medical students' education. *Philosophy, ethics, and humanities in medicine*, [e-journal] 3(1). Available at https://peh-med.biomedcentral.com/articles/10.1186/1747-5341-3-10 [Accessed 15 July 2016].

Shaywitz, David, 2012. Humanism in digital health: do we have to sacrifice personal connections as we improve efficiency. *The Atlantic*, [online] 31 October. Available at: http://www.theatlantic.com/health/archive/2012/10/humanism-in-digital-health-do-we-have-to-sacrifice-personal-connections-as-we-improve-efficiency/264325/ [Accessed 19 July 2016].

Shlain, Jordan, 2016. Digital empathy . . . cracking the code. [presentation] At: Health 2.0 Meetup, San Francisco, CA, 8 March 2016.

Stingl, Alexander I., 2014. Digital fairground – the virtualization of health, illness, and the experience of 'becoming a patient' as a problem of political ontology and social justice. In: Harry F. Dahms, ed., *Mediations of social life in the 21st century*. Bingley, UK: Emerald Insight. pp.53–92.

Topol, Eric, 2012. *The creative destruction of medicine: how the digital revolution will create better health care.* New York: Basic Books.

Wanshel, Elyse, 2016. Virtual reality lets med students experience what its like to be 74. *The Huffington Post*, [online] 3 June. Available at: http://www.huffingtonpost.com/entry/we-are-alfred-embodied-labs-carrie-shaw-virtual-reality-medical-students-elderly-geriatric-care_us_57505bbce4b0c3752dccbeaa [Accessed 12 July 2016].

Webster, Andrew, 2007. *Health, technology and society: a sociological critique.* New York: Palgrave Macmillan.

Yamin, Cyrus, Emani, Srinivas, Williams, Deborah, Lipsitz, Stuart, Karson, Andrew, Wald, Jonathan and Bates, David, 2011. The digital divide in adoption and use of a personal health record. *Archives of Internal Medicine,* 171(6), pp.568–74.

Zeeman, Laeticia, Aranda, Kay and Grant, Alec, 2014. Queer challenges to evidence-based practice. *Nursing Inquiry,* 21(2), pp.101–11.

12 To hold and be held

Engaging with suffering at end of life through a consideration of personal writing

Kelsey B. Hanrahan

In this chapter I draw on my experiences – through both my written notes and my memories of this time – while conducting my dissertation research in a small rural Konkomba village in Northern Region, Ghana (Hanrahan, 2015). The focus of the research was on intergenerational relationships with respect to everyday practices making up livelihoods. Practices of care are woven into these everyday activities and are an integral part of these relationships. The Konkomba use the word *joo* to express the acts of caring for others. *Joo* can be translated as 'to hold, carry in hand' (Langdon and Breeze, n.d.) or 'to handle' (used by the field assistant providing *in situ* interpretation during interviews). *Joo* conveys not only a form of action, but also a way of being in the world defined by a caring connectedness – a characteristic possessed by a person with a general disposition to consider the needs of others. To handle someone is therefore to participate in the life of another. Although care is part of everyday life with responsibilities embedded within particular intergenerational relationships, intense moments of dependency such as end-of-life illness are moments of intimacy that offer insight into the depths of interpersonal connection.

I was present for one woman's death. As Nyaa Uchain was dying, she invited me to hold her and her family welcomed me to handle her. I see these invitations as gifts – to be present for, and participate in, an intensely intimate period in her life. I revisit these experiences in this chapter through the act of writing, which included both the in-field documentation of observations and conversations and the exploration of personal thoughts and feelings, as well as retrospective considerations of those writings. I discuss the methodological impacts of these experiences and how they came to shape my approach to conducting fieldwork, building personal connections and understanding the community in which I lived, worked and cared.

Nyaa Uchain was a tiny, quiet woman with a gentle way about her. She liked to sit with others, sometimes telling stories about her husband. She described him as having loved her and handled her well long before they married, when she was still a girl. When she became ill, I would visit her daily. I started to help financially – with hospital fees, medicines and purchasing particular foods. I was also invited to participate in discussions regarding her care and the decision-making process concerning her hospitalization. And while this story may seem to begin when I first met her, or perhaps when I first started helping her family provide care, after reflection I am inclined to suggest that this story begins in many moments beyond my initial meeting and financial support. It begins with her head resting on my lap, with a look between my eyes and hers. It begins when I visit her alone and find that it does not matter that we do not speak the same language. It begins when I ache with the inability to ease her pain. The experience of being with her as she was dying was a struggle – the struggle to comprehend the experiences of another. Yet as I engaged with these experiences, the story began to take shape for me. It has become a story made up of moments of suffering and moments of connection.

Handling Nyaa Uchain

In 2011 I first met Nyaa Uchain and became friends with her and her family. I would often visit the compound to greet Nyaa Uchain and I also got to know her daughter; we were both shy talking with each other, but we would sit together as I would play with the baby. Hers was a busy compound – her home included her son, his wife and their children, her daughter and her infant child, her husband's other widow, along with a young woman who was staying in a room with her three young children.

By the start of the rains in 2012, Nyaa Uchain had fallen critically ill and her children's concerns were palpable. Nyaa Uchain's family admitted her to the local hospital and subsequently had her transferred to the regional capital for more care. She returned home, however, after resisting treatment. In the few months that followed, Nyaa Uchain was confined to her room. She had all but stopped eating and drinking, having just fleeting moments of interest in eating porridge or drinking pito. She grew very weak and had difficulty supporting herself in a seated position or shifting her position while she was lying down.

At this moment in her life, a community handled Nyaa Uchain. Her daughter left her boyfriend's home to live with her mother and to care for her. She had her infant son with her, and as her mother's condition worsened, the daughter embodied her worry. She ran out of breast milk to feed her child, growing thin herself as she focused her physical and emotional energy into her mother. Nyaa Uchain's daughter-in-law often stayed home to provide the intimate care that was needed more and more each day. Her son, as head of the household, was considered to be the primary decision-maker with regards to his mother's care. He was in charge of making financial arrangements for her hospitalizations and pharmaceutical needs, as well as accessing potentially useful herbal remedies. He was supported by Nyaa Uchain's co-wives' sons, who managed many of the practical affairs, such as accompanying her to the hospital.

As her condition worsened, the community of care around her grew. Her children spent more and more time with her in the home, neglecting their farms at a critical time of year when many crops were being planted and would serve to feed the family for the coming year. Her daughter and daughter-in-law were always nearby and women from outside the household supported them. An elderly woman, Nyaa Ntibi, visited her every day. She came to visit and provide comfort, as well as to show the daughters how to care properly for the dying. Another woman left her own room late every night to tend to her throughout the night. The breadth of Nyaa Uchain's relationships was revealed as extended family and friends provided this support, as well as providing money for hospitalization, medications and herbal treatments. Still others dropped by to pay a visit, to bring small gifts of food and drink, to ensure she was comfortable and to provide advice and encouragement to her family.

An invitation to hold and handle

In relationships of care, one may recognize suffering and work to acknowledge and attend to the needs of another. In this way, Nyaa Uchain was well cared for. In speaking with other women in the village, they would describe to me how well the daughter and daughter-in-law were handling Nyaa Uchain – she was regularly bathed and dressed in clean clothes; they cared for her sores; they tried to get her to eat and drink. And yet while Nyaa Uchain was undoubtedly experiencing physical discomforts associated with her diminishing health, in speaking with me, her daughters emphasized the suffering they could not address. I learned about her anguish caused by her inability to hold her grandchildren and care for them. Her daughters reported that she often cried throughout the nights:

> [S]he's going to die and leave us soon. [My mother] cries because she was supposed to be healthy and carry the young children in the house so women could go to farm. When she looks at children she cries and wants to be help-ing. [She says,] 'I've met my grandchildren, but I can't handle [take care of] them.' (Nyaa Uchain's daughter, 2012)

More than once I came to greet her, and she was upset because she could not hold her grandchildren. Not only was she physically weak, but traditions dictated that her condition should not be exposed to small children. Her daughter was therefore incapable of staying with her during the night because she needed to be sleeping with her smallest child. Nyaa Uchain's condition was breaking her connections to others in the community, and was a source of her suffering that could not be tended to in the ways her physical needs were.

Suffering is relational, 'transcend[ing] the space of the body through rela-tionships that become established through it' (Olson 2014, p.431). Suffering is shaped through relationships with others – how it is communicated, understood and addressed – and relationships in turn are created through this engagement with suffering.

> Uchain gestured for me to come to her, so I sat right down beside her mat and held her hand. She would run her hands all over my head and face, try-ing to talk, but only making sounds, although at one point she was saying how I love her, I love her, and she called me her child. . . . It was all quite

overwhelming, like she wanted me to help, but there was nothing more I could do. We sat with her about an hour, until she fell asleep – she had started to cry at one point because she thought I was leaving and crying out not to leave her, so we stayed until she fell asleep. (Hanrahan, 2012a)

Nyaa Uchain's eyes, her hands and – while she could still speak – her words begged for my presence. Although throughout our relationship she had always simply called me Ukalnpii ('white lady'), when she fell ill she also began to call me her child and her daughter. While I have seen others do this – invoke a form of kinship relationship to emphasize or impose certain connections or responsibilities – it was an incredible moment for me. In this way, Nyaa Uchain brought me into an intimate space with her.

At the same time that she was kept from holding children, Nyaa Uchain sought to be held. In this community, touch is infrequent. Babies and small children may be held, kissed and comforted with touch but among adults it is rare to witness touch. I was becoming increasingly aware of this cultural norm and in Nyaa Uchain's room touch seemed limited to functional interactions.

At one point, Uchain maneuvered her head into my lap – she [is weak] and moves jerkily – in big, quick bursts of efforts. [Nyaa Ntibi] helped me lay her back down on the mat . . . [The two young women] just watched. I couldn't do much because the angle she was at, and [Nyaa Ntibi] is only so strong. It was like the last couple of days – very happy to see me, reaching for me. Today she also wanted me to lay down beside her, although I kept telling her *lignir* [it's okay]. It's very intense in the room – my interactions with Nyaa Uchain, her daughter watching in surprise. I go and hold her hand because she seems to want the closeness, the physical contact. I don't know if it's common to give, even when it's just family and I don't see anyone else touching her much, except for practical purposes. (Hanrahan, 2012b)

I still struggle to understand whether I should have touched Nyaa Uchain the way I did and whether I should have accepted her head in my lap and reached out for her hand. I continue to wonder what it meant to her and what it meant to her family.

Personal writing and understanding suffering

Although I had previously lost people in my life, it had always been at a distance. Sitting with Nyaa Uchain at the end of her life, I was invited to participate in her dying – to not only witness, but to handle her in this intimate moment in her life. It was as destabilizing as it was enriching. I struggled to understand what it meant to care for someone who was dying and to be invited into the experience. At the time, writing notes was a frustrating process – little of what I felt seemed to find its way onto the page in a way that I felt I could identify with. Through the second process of writing – for the panel discussion and again in the preparation of this chapter, I became aware that the uncertainty was part of the experience.

In re-reading my notes from that period, I can see my own struggle – between the desire to follow impulses of care, responding directly to this woman and the pull of an obligation to follow the cultural norms I was encountering. There are hesitations in my writing; in the excerpt above, I struggle with wanting to provide the physical touch Nyaa Uchain obviously desires, and wanting to respect the family and follow their lead to show me how to handle her. These hesitations remind me that I am working in a liminal space, alternately guided by the expectations of being a guest and being a member of the family. But more than that, the liminal space is also a result of struggling to make connection between two people whose bodies are having very different experiences.

I encountered a new, intense experience for which I was not prepared, and by embracing the relationships I had built within the community I was able to explore new forms of connection. First, I abandoned my dependence on conversations and explanations. I started visiting Nyaa Uchain without my interpreter, alone or sometimes with a friend with whom I could only speak in very simple exchanges. We communicated with gestures and suggestions, and in a very broken, very limited Likpakpaln (Konkomba language). But mostly we just sat with Nyaa Uchain. During these visits, I stopped trying to listen, and I just started feeling. I was not trying to feel what they were feeling, but instead I started to acknowledge my own experience of the situation and the legitimacy and importance of my presence in that room. I started to see the comfort that I brought to Nyaa Uchain when I held her hand and accepted her head laying in my lap and the surprise and happiness expressed by her daughters by those same actions.

I struggled with what seemed the futility of a research plan with data collection methods, wanting to dismiss participant observation as wholly inadequate for understanding intimate experiences. It was through writing and reflecting, back and forth, that I came to understand that I was not there to either observe or participate, but that I was part of the experience for all those involved. My experiences, our experiences, created a shift in my approach to my work. In the months that followed, I engaged in new ways with others; I let go of my expectation that dialogue would mediate our interactions. These experiences and writing about them allowed a deeper understanding of my position in these moments, and an awareness of the importance of the connections I made with Nyaa Uchain and others as a result of being present in these times.

To hold, to handle

Understanding how people support each other in life is inextricable from how people support each other at the end of life and how death can be incorporated into the lives that continue on. Care is demonstrated in action, in the handling of others, and carries with it implications of emotion and value that extend beyond the direct result of action (Tronto, 1993; Bowden, 1997; Kittay, 1999). There was a shift in the home during the months leading to Nyaa Uchain's death. The women focused their labour on the immediate concerns of intimate care for their mother, rarely farming or going to market. Children no longer ran through the compound. Nyaa Uchain's son also struggled to go to farm, admitting that when he was at farm his mind was in the home, on his mother, and he could not farm in that condition. Life was focused on Nyaa Uchain and turned ever-so-slightly to the life that would continue after she passed away.

I also turned towards her; my daily activities and thoughts were permeated by her presence and her needs. The experiences of handling her shaped me as an individual, in turn shaping my presence as a researcher in that community. If someone is suffering, to care for her involves recognizing her needs. In my relationship with Nyaa Uchain, I struggled to accept that while another person's suffering is never fully comprehensible to another (Cassell, 1991; 2004) 'we can, however, learn to recognize the particular purposes, values and aesthetic responses that shape the sense of self whose integrity is threatened by pain' (Cassell, 1991, p.24). Nyaa Uchain longed to connect with others the way she expected to as a mother and grandmother – to hold, to be present. However, at the end of life, she needed to be held and to have others handle her in this moment.

I did not understand her suffering. But I do now understand that I was invited to handle her. I believe I offered Nyaa Uchain something she needed in those moments and I know that in return I was gifted the opportunity to learn about the deep rooted connectedness that can be contained within care. Not only through my in-field writings, but also through the retrospective consideration of those writings, I have gained a greater and more personal appreciation for the ways in which relationships are a foundation to the research process, and that the research process is a deeply personal experience for those involved. The intensity of connections that emerge in moments of intimate experience, and the acts of attending to the needs of another highlight a dimension of care that is more than the everyday experiences and practices of meeting the needs of another. There is the potential for intimate connection to emerge when we are open to care – a connection that can occur when we allow ourselves to welcome relationships within the context of research.

Works cited

Bowden, Peta, 1997. *Caring: gender-sensitive ethics.* New York: Routledge.
Cassell, Eric J., 1991. Recognizing suffering. *Hastings Center Report*, 21(3), pp.24–31.
Cassell, Eric J., 2004. *The nature of suffering and the goals of medicine.* New York: Oxford University Press.
Hanrahan, K., 2012a. Personal visit with Nyaa Uchain. [Notes] (Field notes, 15 April 2012).
Hanrahan, K., 2012b. Personal visit with Nyaa Uchain. [Notes] (Field notes, 17 April 2012).

Hanrahan, Kelsey, 2015. *Living care-fully: labor, love and suffering and the geographies of intergenerational care in northern Ghana*. PhD. University of Kentucky.

Kittay, Eva Feder, 1999. *Love's labor: essays on women, equality, and dependency*. New York: Routledge.

Langdon, Margaret A. and Breeze, Mary J., eds., n.d. *Konkomba-English Likaln-Likpakpaln Dictionary*. Tamale: Ghana Institute of Linguistics.

Nyaa Uchain's daughter, 2012. *Discussion of end of life care*. Interviewed by Kelsey Hanrahan and Ntesi B. (interpreter). [Interview] Binalobdo, Northern Region, Ghana, 12 April 2012.

Olson, Elizabeth, 2014. Ethics. In: Roger Lee, Noel Castree, Rob Kitchin, Victoria Lawson, Anssi Paasi, Chris Philo, Sarah Radcliffe, Susan M. Roberts and Charles W.J. Withers, eds., *The Sage handbook of human geography*. London: Sage Publications. pp.423–44.

Tronto, Joan, 1993. *Moral boundaries: a political argument for an ethic of care*. New York: Routledge.

13 Inhabiting research, accessing intimacy, becoming collective

Karen Falconer Al-Hindi, Pamela Moss, Leslie Kern and Roberta Hawkins

Note to the reader

References to 'she' and 'her' in this chapter are deliberately ambiguous, as this is part of our becoming collective. The reader is invited to join us in suspending the usual correspondence between noun and pronoun. We hope you find our use of language productive of new thoughts and ideas.

Why can't I be part of your memory?

Together, we investigate joy in academic spaces as feminist geographers through collective biography. Investigating joy is our response to restrictive and compromising academic labour conditions by affirming the radical potential of becoming joyful subjects in and through our work (Kern, et al., 2014b). Collective biography is a collaborative methodology used by feminists to examine the processes through which subjects emerge (Haug, et al., 1999; Davies and Gannon, 2006). Our collective biography work entails annual writing retreats and twice monthly Skype videoconferences. During the retreats, we generate texts from our individual memories of formative moments. By formative, we mean those experiences or events that feel pivotal to the topic we have chosen to investigate. Once the texts are written, group members collectively discuss and analyse them. After a couple of rounds of revision, we complete our writings of the moments, name them, and file them as 'Final Memories'. These texts then become the basis for scrutinizing the way in which we became subjects as well as the way in which we continue becoming subjects (Davies and Gannon, 2006, p.7).

Working with joy led us to search within the feminist geography literature for scholarship that mentioned or conceptualized joy. We were surprised to find few feminist geographers working with joy, but interestingly the writing that did allude to joy also involved intimacy. For example, Lees and Longhurst (1995) recall joyful memories of an intimate feminist geography gathering in Aotearoa/ New Zealand and Boyer (2012) notes joy as one affective dimension of the intimate experience of breastfeeding in public. Although elucidating the connections between joy and intimacy was not originally part of our project, the literature indicated that these links might be worth pursuing. Feminist geographers produce accounts of intimacy in keeping with a long tradition of bringing into view spaces, places and experiences that have been deemed private, innermost and even mundane, such as the home, the family, the body and emotions (e.g. Bondi, 1998; Elwood, 2000; Bondi, Davidson and Smith, 2005; Dyck, 2005). Such insights have fuelled engagement with the notion of the global intimate (e.g. Mountz and Hyndman, 2006; Pratt and Rosner, 2012; Brickell, 2014; Pain and Staeheli, 2014). As well, intimacy has been central to feminist research via sensitive readings of power through social positionings and connections in encounters with research subjects and participants (e.g. Cupples, 2002; Valentine, 2002; Smith, 2016). Following these rich and generative accounts of intimacy, we decided to explore its connections with collective biography.

Our entry into a discussion of intimacy in collective biography is the methodological implications of the uses (and perhaps abuses) of intimacy in research. Collective biography requires both rapport among and a long-term commitment from group members. In setting up a collective biography workshop, Davies and Gannon (2006, pp.8–9, 135 and 137–40) suggest a week-long retreat is sufficient for creating close relationships in a short period in order to provide spaces to negotiate vulnerabilities. We found that such a time frame only allows for rapport in the sense of ease and camaraderie, rather than rapport as a deep mutual understanding. Indeed, conflicts and breakdowns in collective biography groups can often be traced to short timelines and scenarios for rapport-building. In our experience, the relationships needed to be more extensive and deeper than those arising from the process of colleagues doing research together and they required much more time and effort to develop. Moreover, they required an evenness of commitment and participation, as well as conditions where academic hierarchies and power imbalances are part of ongoing discussions. In writing up the analysis of the memories generated in a workshop, Davies and Gannon (2006, pp.129–32, 136) note that the long timeline for writing and publishing coupled with placing responsibility for writing first drafts on one person can cause tension within a group. Davies and Gannon's workshops also had varied membership, including standing members, students taking their courses and visiting colleagues. In contrast, our collective biography group remains small and unchanging. Because we are all professors, we participate in, and acquiesce to, academic writing and publishing processes and timelines. Whereas the memories generated in Davies and Gannon's workshops could be used by any group member for analysis beyond the initial manuscript, we chose from the beginning to analyse and write up all analyses collaboratively. This even participation and commitment to the research over a long period of time deepened our familiarity with and appreciation for each other, thus generating the space for intimacy to emerge.

At the beginning of our work, while each of us committed to a long-term collaboration, we really did not know each other very well. Karen and Pamela had written together before, but were not very close. Leslie and Roberta had written an article on which Pamela had commented. Over time, familiarity with each other and sensitivity to each other's experiences supported our efforts as co-researchers working and writing together. What initially might have been described as congeniality gradually transformed as we found ourselves moving away from the notion that we were a close group of four colleagues working together on a project, towards the sense that we were a collective generating multiple projects. Even though we knew that spending research time together would not necessarily lead either to a collective or to intimate relationships, we wondered what it was about becoming collective through the practice of collective biography that created the possibility of intimacy during the time that we did spend together.

In reflecting on this transformation over the past few years, we often wondered how the shift took place, the shift from collegial co-researchers to a collective with rapport in a deep sense, having close and harmonious relationships with well understood feelings and good communication. We recognized that we were doing collective memory work, we were co-writing papers, and we were co-presenting at academic conferences. Yet we still attributed ideas to one another and referred to each other through tracked changes in our documents. While some of us were ready to become a *we*, there were still four *I's* working together. Whenever we spoke of collectivity in some form, our conversations would often return to an incident early on in our collaboration where one of us was so immersed in another's memory that she felt she was there, and was remembering the experience herself. In response to the prompt at our first retreat, 'Do you remember a moment of joy while in a collaborative endeavour?' she wrote about the beginning of a longitudinal project with colleagues that would include video. She remembered looking online for cameras but was adamant they did not buy one; she was sure they had bought one, right then, right there, for it made perfect sense for the story.

> She reads her memory. Something about studying on a Sunday afternoon in a coffee shop. I could almost feel the backpack against my leg.
>
> Warm air, a hazy hue of yellow from the walls, the smells of coffee and baking and Elsa and Ana and I crammed together in a booth working and laughing with ease . . . and then . . . [. . .] I'm wrenched away from the comfort of the coffee shop to another room.
>
> Wait. What? She doesn't? She didn't buy the camera? I act emboldened, perhaps too much for not knowing her all that well. 'What? She didn't buy it? Why not? You had internet access, too, right?'
>
> 'What is she doing?!' [. . .], the very idea drawing my eyebrows up and jutting my chin forward. Did she plan this ahead of time? Had she been waiting for the right moment, the sharing of the right sort of memory, to unleash this truth correspondence issue?
>
> These new colleagues, near strangers, ask me questions and I bite my lips slightly, press my thumbs and fingers together clenching.
>
> Teasing her a little more, mimicking writing in the air with my pen, [. . .] 'Of course she bought it, you're just remembering wrong.'
>
> I defend the memory. It's mine. I know how it happened.

Like the tuber in a game of 'hot potato', her suggestion bounced among us.

I squeeze my eyes and nose trying to think harder. I need to open up, to loosen up. Can I be sure of any of this? I feel crowded and uncomfortable. How can we all fit into the booth in my memory?

'But she didn't buy it.' I hear off to the right, someone saying, 'But does it matter?' And then more banter about authenticity, genuineness and truth. Friendly banter? I think so.

We initially understood the intrusion of one of us in another's moment as an analytical point, as concerning the epistemological status of truth as it has been problematized by the feminist methodological literature: does it matter how an event *really* happened, or is it a question about knowledge production? Feminists challenged the idea that there is a singular, unchanging truth that explains the social world (Hekman, 1997; see also van Zoonen, 2012). Refusing the idea that knowledge is a thing to acquire, feminists turned to questioning what is known, who is the knower, and how one comes to know (e.g. Haraway, 1988; Hawkesworth, 1990; see also Hekman, 2010). Later, we realized that there were additional layers of meaning to this interaction that went beyond the question of how truth is produced (Foucault, 1997; Braidotti, 2011); these layers brought us towards the realization that through collective biography we might come to inhabit one another's memories. Inhabitation was an unexpected dimension of the methodology, and thus came to occupy an important place in our reflections on our research process. The collectivity of memory work, mobilized around and through revision and analysis, entails the group working together to clarify the written account of each memory and then to write up the analysis of the final memories in terms of the concept or question at hand (in our case, joyfulness in academic practice). Yet here we were *actually* making the memories collective as we began inhabiting part of all the memories.

Colebrook (2000, p.3) writes about the distinction between interpreting a text, and inhabiting a text in the context of Deleuze's writing:

> Rather than seek the good sense of a work, a Deleuzean reading looks at what a philosophical text creates. To see a text in this way means abandoning the interpretive comportment, in which the meaning of a text would be disclosed. In contrast, one *inhabits* a text: set up shop, follow its movements, trace its steps and discover it as a field of singularities (effects that cannot be subordinated to some pre-given identity of meaning). (emphasis in original)

With our collective biography of joy in academic practice, following from Colebrook, we inhabited the final memories in the way that researchers analyse texts (see Kern, et al., 2014a). Yet in our memory work about becoming collective, we inhabited not the texts but the memories themselves. Our memories, once shared, permitted each of us to take up space and insert ourselves into each other's memories methodologically; not as an act of colonization (acts that have historically erased minoritized groups' lives, see Solórzano and Yosso, 2002; Chapman, 2005), but as a response to being invited to join into a memory of an experience replete with bodies, affect and emotion. As part of our work on becoming collective, we forgot about what meaning the experience in the coffee shop held for her in terms of joy, and began tracing what the memory did for us as a group. We followed its movements in different directions, ones that did not resemble her own recounting. Our comments pointed towards other possibilities that could have been taken up but were not. Our continuing discussion beyond our first retreat of that final memory traces an effect of that text outside the task of finding joy as an academic in an increasingly restrictive institution. That effect, in a sea of other potential effects, has disclosed a previously obscured aspect of the process of becoming, transforming us from *I's* to *we*.

Inhabiting memories as a pathway to generate intimacy has not always been a comfortable or seamless process. Inhabitation leading to intimacy is an uneven process in terms of both time and intensity, a murky route with its own set of effects. Our interactions have been intense, rife with an array of complex emotions. Uneasy self-reprimands of a latent positivism around ideas of what really happened lay alongside the troubling notion that messing with someone else's memory might be a kind of ontological infringement:

> As the months passed she seemed willing to entertain her version, so over time my hold on the original memory relaxed also. Ignoring the thin wire vibrating between my shoulders whenever the issue arises, I have remained silent, trying to blend in with the group on the issue of the camera memory. I think I've succeeded.

In our efforts to think through how this transformation from *I's* to *we* came about, we engaged in a collective biography via Skype. We posed questions to elicit formative experiences to write about. The final memories produced include the one excerpted above. This exercise suggested to us that there are sundry and unexpected paths to intimacy arising out of a turbulent set of conditions and emotions that generated the potential for becoming collective (Kern, et al., 2014a). In the rest of this chapter, we work from the texts of our final memories to explore in more detail how inhabitation captures our collaborative research process. We follow with some thoughts about differences between investigating intimacy and generating it.

Access routes to intimacy

Since our second writing retreat (held at a cabin in a Nebraska state park in 2014) we have often recollected one particular event – an encounter with several four-legged creatures – through stories, jokes and even teasing among us. The mice incident, as we have come to call it, happened near the beginning of the retreat and was among the unforgettable memories from our week of collaboration. Given its status as a shared, memorable experience from our time together, we decided to use collective biography to generate data from this event. While we had all agreed that the mice incident was part of the pathway to intimacy for our group, we were surprised at the variation in our experiences. What emerges from our initial tracing of the effects of these final memories is that the generation of intimacy happens via multiple, unpredictable and sometimes discomforting access routes. These routes were turbulent, involving a cocktail of emotions that were not typically pleasant. We each embodied the experience in ways that produced diverse feelings of anxiety, vulnerability, disappointment and relief. Despite our differing, unsettling experiences, our accounts point to the relational nature of our access routes and shed light on becoming collective.

> Shivers run up and down my spine – it may as well be the mice crawling on me. [. . .] My heart is hammering. I glance quickly to the corners of my bedroom, my eyes dart around the carpeted floor – they could be in here – they could be anywhere. An icky feeling in my throat, I try to stay focused on this task. Empty and repack the suitcase. Get ready to leave. [. . .] I go into the living room and ask for help. I know they are only mice and all the same, I know I don't want to do this alone.
>
> She turned to face me as I began to poke gingerly around the frayed-edged hole in the side. 'Eww', she said, 'it's so gross.' I unzipped suitcase pockets and patted them down, feeling for tiny, live bodies. My shoulders unclenched a little when the lumps under my touch turned out to be cracker crumbs, shredded wrappers and cookie remains. [. . .] My gratitude for the chance to help flowed into the urgent movements of my gloved fingers. [. . .] I felt, rather than heard her say, something that I took up as embarrassment. Perhaps it was her bright pink face that raised a question in my mind. The question drew my vision away from the goodies toward the other contents of the suitcase pocket: her lingerie. I'd been pawing through her undergarments, as they had been together with the treats.

In these final memories two of us share an experience when one asks for help checking her suitcase for mice. The discovery that the mice have been in the same pocket as her underwear leads to embarrassment generated by the forced intimacy of having your underwear pawed through, or 'pawing through' someone else's. Physical proximity to personal items in a moment of distress generates embodied routes to intimacy. The suitcase owner experiences vulnerability through her fear of the mice, needing to ask for help, and having her underwear exposed. Her sheer relief that she was not dealing with the mice alone and would soon be able to leave the infested cabin for a new one is part of a cocktail of emotions that eventually, albeit uncertainly, generates intimacy. In another's experience the idea of moving cabins to escape the rodents does not initially sit well, putting her at odds with the others.

> It's now clear that we're moving to another cabin, different than this one, hopefully without mice but surely not as perfect. My throat is a little constricted and I hold my stomach tight. I swallow. I release my lip from my

teeth and take a last swig of the now-lukewarm coffee. They're all in motion now, chatting all at once, voices high with relief. I smile and begin to nod, then stand up and move to the kitchen. Dumping my cup out into the sink, I grab a canvas bag and start packing.

Her emotions, expressed through subtle bodily actions (e.g. tight stomach muscles), first centre on sadness and anxiety about having to move and perhaps losing the perfection and burgeoning intimacy she had identified with her experiences in this cabin. As the rest of the group prepares to leave, her emotions then shift to absorb the relief and excitement filling the cabin. She moves to help with the preparations. This cocktail of emotions illustrates that intimacy is not necessarily felt as joyful or positive for those involved. It was the intimacy formed through writing in the perfect cabin all day that led her to feel anxious about the move but the shared and relational feelings of relief and excitement ultimately encourage her to move. This point is illustrated even more prominently in another's experience of the mice incident later that night:

> How could have I shrieked so loud? On a chair, too. At least she was on a chair. And her, what a trooper. I'm sure she feels bad about the cabin. I'm just so tired. Maybe that's what it's about. Maybe if I can sleep tonight, I'll be okay. I'll get rid of the nausea, the headache, and the embarrassment. [. . .] Tomorrow should be better. No mice. If I can sleep. I pull my knees up and place my right foot on top of my left. Do they think that I'm odd? I wonder if they will still think that I have something to contribute. [. . .] I make it to the bathroom. . . . I lean over the sink, tightly gripping the edge of the counter. I let the cold water drip from my face instead of using a towel. It is going to be a long night.

For her the unexpected exposure of her vulnerability brought on by scurrying mice and her embarrassment and concern over her reaction to them are intensely embodied. While this experience generates intimacy among group members, the process is not pleasant and is in fact infused with aspects of insecurity, doubt, and dread about how future work amongst the group will go after her vulnerability has been exposed in such an awkward way.

In all of these excerpts intimacy is generated within the group in relation to other group members and their feelings, experiences and actions. It is significant that over time, a distressing incident became a shared source of humour. Indeed, we find that many memories of becoming collective involve the emergence of humour out of anxiety, concern and fear. We speculate that the turbulence itself – as an integral part of the rocky access routes to intimacy – is key to our ongoing process of becoming collective, in that we have moved towards collectivity through meaningful experiences of connection that have not simply involved polite collaborative exchanges, but rather challenging, embarrassing and stressful events. In tracing the effects of our stories about the mice incident, we are in no way prescribing how intimacy can be achieved in a collective writing group but instead we are trying to follow how this worked through one event experienced by our group. Importantly, we acknowledge that these experiences could easily have led to feelings of alienation rather than intimacy because of any number of issues, including unequal power dynamics or professional hierarchies. Time matters, too: the mice incident happened after two years of working together and as such we were already navigating access routes to intimacy and developing collective connections.

Becoming collective

Many academics have written about collaborative writing methods. For example, Wyatt and colleagues (2014a) referred to themselves as assemblages by combining the initials of their first names, such as, JKSB, JKB, JB, and B. Davies (B) engaged the notion that writing together is usefully thought of as rhizomatic in ways that resonate for us:

> Mangroves, for example, live in water, and their roots rise up into the air [. . .]. [T]hese beautiful and chaotic trees come closer to what we did [. . .]. We leaned and we lurched. We put unbearable weight on each other. [. . .] Which tree was which – the question no longer made sense. (Wyatt, et al., 2014b, p.409)

For us, questions about who wrote or said what no longer make sense as we inhabit one another's memories, think about joy and write together. Our collective is committed for the long term and this has contributed to our becoming intimate: since the beginning we have neither added nor lost members. Our inhabitation of collectivity contrasts with JKSB, who form multiple assemblages, coming together around one or two writing projects. Given our group's longevity, the mangrove metaphor may fit us even better than it does JKSB. Our discussion of the memories and the analysis of the final memories show that inhabiting is part of our research. The interactions we have around resurrecting memories that had settled in our bodies and the way in which we make sense of the content of the memories shape how we relate as a group. Some of these interactions reflected purposeful decisions (eliminating the use of tracked changes in our manuscripts), while others were simply doing the work (conversing over Skype about the research), and still others were incidental (attending to a family member's medical needs). In what follows, our analysis of excerpted final memories illustrates different dimensions of our inhabitations as part of our becoming collective.

> I hold my hand in front of the vent letting the cool air pass through my fingers. Rather than turn to speak, I simply call out, 'No, I didn't go to that session, tell me more.' As I listen to the details, I gaze along the roadside. [. . .] I watch the cars on the other side of the road pass by, then pull my eyes back to the white lines separating the lanes. She announces: 'It's not that

much further.' I look up, finding a wooden structure to the right, not too far off in the distance. 'Is that where we're going?' The answer is short: 'Yeah.' Leaning forward, letting the air from the vent cascade around my face, I close my eyes.

There is a sense that the writer is cared for: someone else is driving and knows the way. She can relax, rest her eyes, and take in the cool air. It seems that the others agreed, tacitly, to give her the break she needed. It is clear that we do not all have to be great at everything, or responsible for everything, or engaged all the time, or even express everything aloud. The group made their way towards the retreat destination despite her need to relax. An unspoken ability to respect one another's needs yet proceed together is an effect of the intimacy gained via various access routes, and allows us to continue becoming collective.

The history of our collective is rife with pleasure and humour even around unfortunate incidents. Many of the final memories written for this chapter are moments that we often return to as a group, fondly reminiscing and laughing about the details despite some of our initial interpretations of the experiences as unpleasant. Below, initial concern for one group member moves into good-natured teasing as it becomes clear that she is unhurt.

She throws open the van door and the smell of gasoline rushes in. Something about the pump spurting out gas . . . her khaki pants are soaked through. I'm leaning out the window now, 'Are you ok? What happened? Oh my god.' Concern slowly gives way to relief and I let out a stifled giggle. She's fine, just wet and shocked and incredulous. I twist back down into my seat. [. . .] It's okay to laugh now, even tease as she uses the van for cover and strips off the smelly pants without hesitation. At the cabin we hang them up to dry on the front porch, a limp and pungent flag to mark the start of the new retreat.

Although *her* pants were soaked with gasoline, *we* adopted them as our flag. A similar willingness to take a risk with a joke is evident in the excerpt that follows.

My chest glows with the knowledge of a present well chosen. She opens the head scratcher box immediately and plunges it up and down on her head right there at the table. She feigns a blissed-out, calm expression. We're all grinning. I feel my cheeks stretching up to my ears. When she's done her hair stands up like a little nest and [we] laugh at the sight of it, so out of place in the middle of the restaurant. Alain and Jeanette chuckle too yet I see them looking around furtively at the other patrons to see that we're not disturbing them. We're wiping the last of the colourful sauces from our plates as Alain announces that it's time for him to take her to the airport. With a quick round of hugs and well wishes, they walk out into the cool damp evening air and are gone.

The warm yet unstable affect generated through our interactions illustrates, for us, becoming collective. All six people around the table appreciate the gift's presentation and ensuing merriment; however, two – Alain and Jeanette – feel a bit outside the circle. We four have a deeper understanding of why the head scratcher is funny: A result of discussion in our memory work. Although there is no effort to exclude Alain and Jeanette we have generated an inside joke. This joke and the head scratcher itself appear over the next few years in our videoconference meetings reigniting the initial humour, affection, and feelings of intimacy.

This memory of mini golf also concerns a gift, albeit of a different kind:

> The blue barrel with the giant duck [. . .] took me five tries – ugh. 'On three!' cajoled someone with a camera, and I looked up, mustering a smile. Despite the meticulous score-keeping in my tiny notebook, we moved as one through the 18 holes, uncharacteristically silent save for gentle teasing about missed pars and water hazards. [. . .] As we neared the end of the mini golf course one of us managed a hole-in-one. She jumped up and down, gleeful. Congratulatory sounds came from the rest of us; I felt her obvious pleasure in my shoulders (looser) and my face (smiling). 'Thanks! Thanks!' she said. 'But no,' she offered, '*we* got a hole in one.'

At the start of this excerpt, she hints that the others have dragged her along to play mini-golf. But by the end, she relates the feelings of belonging to becoming collective, through both her generosity in sharing the hole-in-one and the intimacy of feeling her pleasure in her shoulders. Collectivity and its affect are elusive and can be unexpected, something felt in sighs and flashes between the recognition of individual bodies and largely separate lives.

Our final memories of becoming collective illustrate the sometimes fleeting and unpredictable ways in which this becoming happens. And these are not necessarily through our purposeful attempts to create a *we*. These conversations likely helped with our group dynamics through sharing experiences, some mundane (driving), some planned (dinner), and some unexpected (gas spill). Yet it was affect generated through unscripted moments, both joyful and distressing, that facilitated our movement. These moments and our reactions to them enhanced intimacy within our group and facilitated a process of becoming collective.

Reflections on writing intimately

Collective biography asks us to consider data creation and analysis as inseparable and entwined in our research process. The initial draft of each memory is only the beginning of data creation; rounds of discussion and revision transform each into what we call final memories, or 'analytical data' (Hawkins, et al., 2016). While the initial drafts are clearly generated by individuals, subsequently the memories cease to belong to each of us as individual researchers, and become shared data. This happens through the deliberate practices of discussion, revision, and naming of the final memories that are key to collective biography. For us, the deep engagement with this analytical process led us to not just share the final memories as data, but also to inhabit them. We noted this earlier when her memory of having brainstormed a new project with colleagues during which they considered the purchase of a camera was so real as to be felt in the body of another. This inhabitation permitted scrutiny of intimacy as a topic while, at the same time, ushering the *I's* towards becoming *we*.

As a collective we fashioned a collaborative writing process grounded in our growing intimacy; nonetheless, there is a difference between investigating intimacy and generating it. Our investigation into intimacy was guided by a collective biography methodology that draws on the centrality of memory in the formation of the subject. The collective biography approach involved deliberate steps and acts that allowed us to analyse the final memories, and this process made inhabiting the memories possible (although this outcome is not certain). In contrast, generating intimacy was less the result of a deliberate set of acts, and more an unpredictable outcome emerging from a combination of moments and practices drawing us together through affection, generosity and the possibility of connections. Time spent working together, even through collective biography, is not sufficient on its own to generate intimacy; rather, intimacy emerges through openness to positive affect. We have found that the turbulent pathways to joy and unpredictable cocktail of emotions from which joy emerges (Kern, et al., 2014b) are not dissimilar to the ways in which intimacy has arisen – both organically and intentionally – in our collective biography work and writing.

Our work in this chapter has been to reflect upon and explore in depth the process through which we are becoming collective. Inhabiting our memories was an unexpected element of collective biography, but we have come to understand this as both generative and as an effect of closeness, familiarity and care. We also understand intimacy among us as developing along uncertain and turbulent pathways that we call access routes to intimacy. The combination of purposeful and serendipitous moments of enacting *we* (becoming mangrove) illustrate the ongoing process of becoming collective. Although we cannot provide a map for others, we hope that our efforts to share our process inspire in readers the same excitement to explore, share and enact intimacy collectively.

We're working hard to stick to our script, as our lines race by faster than they seemed to in rehearsal. My tongue sticks to the top of my mouth, but I can't take a sip of water. [. . .] It's too late now, just keep up. The audience seems to follow along as the dialogue bounces across our semi-circle. 'I really think she bought the camera right then', says one of us, followed by the well-rehearsed rejoinder, 'She didn't, I'm sure. It's not true!' 'Does it matter?' comes the response. My turn now, in unison with the group: 'Yes!' I shout, while others simultaneously shout 'No!' The audience chuckles at our deliberate confusion. We take a few quick breaths, and finally catch each other's eyes. On we go.

Works cited

Bondi, Liz, 1998. Gender, class, and urban space: public and private space in contemporary urban landscapes. *Urban Geography*, 19(2), pp.160–85.

Bondi, Liz, Davidson, Joyce and Smith, Mick, 2005. Introduction: geography's 'emotional turn'. In: Joyce Davidson, Liz Bondi and Mick Smith, eds., *Emotional geographies*, Guildford: Ashgate. pp.1–18.

Boyer, Kate, 2012. Affect, corporeality and the limits of belonging: breastfeeding in public in the contemporary UK. *Health and Place*, 18, pp. 552–60 [Accessed 17 October 2014].

Braidotti, Rosi, 2011. *Nomadic subjects: embodiment and sexual difference in contemporary feminist theory*. New York: Columbia University Press.

Brickell, Katherine, 2014. 'The whole world is watching': intimate geopolitics of forced eviction and women's activism in Cambodia. *Annals of the Association of American Geographers*, 104(6), pp.1256–72.

Chapman, Thandeka K., 2005. Expressions of 'voice' in portraiture. *Qualitative Inquiry*, 11(1), pp.27–51.

Colebrook, Claire, 2000. Introduction. In: Ian Buchanan and Claire Colebrook, eds., *Deleuze and feminist theory*. Edinburgh: Edinburgh University Press. pp.1–17.

Cupples, Julie, 2002. The field as a landscape of desire: sex and sexuality in geographical fieldwork. *Area*, 34, pp.382–90.

Davies, Bronwyn and Gannon, Susanne, eds., 2006. *Doing collective biography*. London: Open University Press.

Dyck, Isabel, 2005. Feminist geography, the 'everyday', and local–global relations: hidden spaces of place-making. *Canadian Geographer*, 49(3), pp.233–43.

Elwood, Sarah A., 2000. Lesbian living spaces. *Journal of Lesbian Studies*, 4(1), pp.11–27.

Foucault, Michel, 1997. *The politics of truth*. New York: Semiotext(e).

Haraway, Donna, 1988. Situated knowledges: the science questions in feminism and the privilege of partial perspective. *Feminist Studies*, 14(3), pp.575–99.

Haug, Frigga, et al., 1999. *Female sexualization: a collective work of memory*. Translated from German by Erica Carter. New York: Verso.

Hawkesworth, Mary, 1990. *Beyond oppression: feminist theory and political strategy*. New York: Continuum.

Hawkins, Roberta, Falconer Al-Hindi, Karen, Moss, Pamela and Kern, Leslie, 2016. Practicing collective biography. *Geography Compass*, 10(4), pp.165–78. doi:10.1111/gec3.12262

Hekman, Susan, 1997. Truth and method: feminist standpoint theory revisited. *Signs*, 22(2), pp.341–65.

Hekman, Susan, 2010. *The material of knowledge: feminist disclosures*. Bloomington: Indiana University Press.

Kern, Leslie, Falconer Al-Hindi, Karen, Hawkins, Roberta, and Moss, Pamela, 2014a. Practicing collective biography in feminist geography. Paper presentation. *Feminist Geography Conference*, 15–18 May 2014. Omaha, Nebraska.

Kern, Leslie, Hawkins, Roberta, Falconer Al-Hindi, Karen and Moss, Pamela, 2014b. A collective biography of joy in academic practice. *Social and Cultural Geography*, 15(7), pp.834–51.

Lees, Loretta and Longhurst, Robyn, 1995. Feminist geography in Aotearoa/New Zealand: a workshop. *Gender, Place and Culture,* 2, pp. 217–22. doi:http://dx.doi.org/10.1080/09663699550022035 [Accessed 20 October 2014].

Mountz, Alison and Jennifer Hyndman, 2006. Feminist approaches to the global intimate. *Women's Studies Quarterly*, 34(1/2), pp.446–63.

Pain, Rachel and Staeheli, Lynn, 2014. Introduction: intimacy-geopolitics and violence. *Area*, 46(4), pp.344–7.

Pratt, Geraldine and Rosner, Victoria, eds. 2012. *The global and the intimate: feminism in our time*. New York: Columbia University Press.

Smith, Sara, 2016. Intimacy and angst in the field. *Gender, Place and Culture*, 23(1), pp.134–46.

Solórzano, Daniel G. and Yosso, Tara J., 2002. Critical race methodology: counter-storytelling as an analytical framework for education research. *Qualitative Inquiry*, 8(1), pp.23–44.

Valentine, Gill, 2002. People like us: negotiating sameness and difference in the research process. In: Pamela Moss, ed., *Feminist geography in practice: research and methods*. Oxford: Blackwell. pp.116–26.

van Zoonen, Liesbet, 2012. I-pistemology: changing truth claims in popular and political culture. *European Journal of Communication*, 27(1), pp.56–67.

Wyatt, Jonathan, Gale, Ken, Gannon, Susanne and Davies, Bronwyn, 2014a. *Deleuze and collaborative writing: an immanent plane of composition*. New York: Peter Lang.

Wyatt, Jonathan, Gale, Ken, Gannon, Susanne, Davies, Bronwyn, Denzin, Norman K. and St. Pierre, Elizabeth Adams, 2014b. Deleuze and collaborative writing: responding to/ with JKSB. *Cultural Studiesó Critical Methodologies*, 14(4), pp.407–416.

14 Intimacy, animal emotion and empathy

Multispecies intimacy as slow research practice

Kathryn Gillespie

Part of my research into the lives of cows in the dairy industry entailed sitting in farmed animal auction yards and watching animals being sold off in rapid succession. During one of these auctions, there was a delay in between the sale of animals. I looked around. I could hear the sound of hooves on the wood ramp leading up to ring, the loud shouts of workers, the bellowing of an adult cow and a high-pitched call of a calf. The large auction ring door opened to reveal the back pens and chutes and a worker struck the cow in the face with the rod he was holding, yelling loudly. The cow refused to enter the ring; her calf was behind her on the ramp and she would not leave him behind. The intention was to sell the cow and calf separately. In order to avoid further delay, the worker herded both the cow and calf into the ring and the auctioneer made an announcement that they would be auctioned separately – the calf first, and then the cow.

The calf was a newborn, no more than a day or two old and his umbilical cord dangled from his belly. He sold immediately for $55. Two of the workers coordinated their efforts: while one distracted the cow, the other opened the exit door just wide enough for the calf to fit through. The calf, startled, trotted through, stumbling at the threshold. The worker standing at the door smacked his rump with the rod he was holding and the calf leapt forward out the door. The door closed and the calf was gone. The cow trotted in circles in the ring and bellowed. From the pens behind the auction ring, the calf called back. After the cow sold for $1,600, they herded her out the door, the door closed, and she was gone.

I could hear the cow and calf continue calling to each other from their separate pens in the rear holding area.

My research is full of these kinds of moments – glimpses into the lives of nonhuman animals in spaces of commodity production. As one dimension of my ethnographic fieldwork, I spent long hours in farmed animal auction yards throughout the Pacific Northwest, in the United States, bearing witness to the highly efficient sale of animals used for dairy, and the gendered commodification of the animal body in these spaces (Gillespie, 2014). Cows and calves are often sold separately from one another; spent cows collapse in the auction ring and in the holding pens behind the auction yard, unable to rise; cows, bulls and steers resist and are struck with rods, shocked with electric prods and sometimes shot (Gillespie, 2016a). As a feminist ethnographer in these spaces, I was deeply moved by the routinized violence of the auction yard (and the dairy industry more generally), and I made a conscious effort to centre these emotional responses of grief and anger as political dimensions in my research (Gillespie, 2016b).

As I observed this commodification of life, I tried to empathize with how the animals themselves might be experiencing the geographic space of the auction yard. I began to see this process of empathizing as a certain kind of *intimacy* with those animals passing through the auctions. At times, the animals' distress was visible and obvious, like the cow and calf whose separation was vocalized before, during and after the sale. And sometimes a cow's body was so worn out, she would collapse in the holding pen, unable to rise. At other times, the embodied experience of the cows was more subtly enacted: their eyes rolled back in their heads, their mouths foamed with saliva or their bodies froze motionless in fear. Almost always, these glimpses of the animals were fleeting (hours at the auction, with some moments barely detectable) – such is the nature of the auction yard. Indeed, the process of commodifying farmed animals increasingly involves the segmentation of industries from the dairy farm, to the breeding farm, to the auction yard, to the slaughterhouse. This segmentation contributes to the literal and figurative distance of consumers from these food production processes and profoundly impacts the (in)visibility of violence (Pachirat, 2011).

What I collected for my ethnographic research was a series of narratives in the form of vignettes of the many animals I encountered. The emotions both embedded in these narratives and my response to them led me to question the role of intimacy in research. Alongside the limitations they revealed about the practicalities of doing research with and understanding nonhuman experience, I queried how intimacy might be understood as a research practice. What can a consideration of intimacy add to multispecies ethnography? How might multispecies intimacy help to theorize intimate research practice more widely?

Intimacy can be understood in a range of ways, illustrated by the diversity of understandings of intimacy in this book. In this chapter, I understand intimacy in two key ways. The first is the intimacy shared among nonhuman animals, demonstrated in the emotional bond shared between the cow and the calf in the vignette above. This bond, as well as the disruption caused by its severance under a system of commodification like the auction, highlights the importance of recognizing the role of emotion and kinship as forms of intimacy that shape lived experience. The second is in the relationship between researcher and research subject. I draw on Gruen's notion of 'entangled empathy' to define intimacy in the research process (Gruen, 2015). This kind of intimacy can be understood as empathy fostered through recognition of multispecies entanglements and their effects. Gruen argues that humans are already entangled in complex relationships of power, care and ambivalent encounter with other species and that recognizing these entanglements is a mode through which one might enact a greater ethic of care in our multispecies social worlds.

This chapter contributes to two ongoing scholarly conversations. I offer a reading of intimacy that extends beyond exclusively human realms and suggests ways of thinking about intimacy in interspecies and more-than-human research contexts. In other words, I make the case for why feminist researchers should consider nonhuman life and how this consideration might enrich scholarship on human and nonhuman social worlds. My discussion also contributes to the literature on multispecies ethnography as an emerging field dedicated to understanding the lifeworlds of a host of different species (Kirksey and Helmreich, 2010). Thus far, multispecies ethnography has not been understood through an explicitly feminist lens. I argue that intimacy as a feminist research practice enriches the field of multispecies ethnography.

In this chapter, I begin with a discussion of the first way I am defining intimacy: as the emotional bonds, ruptures and responses experienced by and among nonhuman beings. In the next section, I argue that attention to these forms of intimacy in other species is a site through which to develop intimacy in feminist research; that is, intimacy between feminist researchers and the subjects they study, whether those subjects are human or nonhuman. The complexities I explore in intimate research practice lead me, in the final section, to consider particular research practices that would foster entangled empathy, those which call for a slower form of scholarship.

Intimacy as animal emotion

In the auction yard, I saw the emotional bonds between animals. I was moved, as a witness to this intimacy, to more carefully consider the ways in which commodification processes for dairy production not only impact the lives and bodies of nonhuman animals, but also shape their social networks and their emotional and psychological experiences of commodification. For feminist scholars attuned to the political function of emotion, recognizing animals' emotional worlds (through, for instance, intimacy produced in these relationships) prompts deeply political questions about the emotional effects of human practices of production and consumption that appropriate animal life in gendered encounters of violence (Gillespie, 2014; 2016b). It is through a feminist attention to the intimacy generated in these emotional encounters witnessed between species that the embodied, emotional consequences of this violence might be better understood. Perhaps the relationship of care, connection and the trauma of separation witnessed between the cow and her calf is a window into considering, with more attentiveness and care, these nonhuman lifeworlds.

Intimacy experienced by members of other species can be understood, in part, through scholarship on animal emotion and cognition. The emotional lives of animals have prompted a fast-growing field of study in which animal behaviourists and ethologists are working to develop a rich literature on the interior lives of a range of species. Many different species (including farmed animals) experience wide-ranging emotions such as joy, love, play, grief, anxiety, embarrassment, fear and empathy (Bekoff, 2000, 2007; Hatkoff, 2009; King, 2013). As humans learn more about the emotional lifeworlds of other species, it is becoming increasingly acknowledged that humans are not the only species that experiences complex emotions and cognition, or develops relationships of intimacy. While intimacy and emotion are not synonymous (intimacy can take forms other than emotion; and emotional response does not require intimacy between two beings), I focus here on the importance of emotion to understanding intimacy in other species to highlight the particular kinds of emotional intimacy I saw enacted among many of the animals I observed. Detailing how emotion is and can be understood in other species is important in order to understand this particular form of intimacy.

Thus, a consideration of intimacy in other species necessitates: (1) an acknowledgement of these varied emotional and cognitive experiences and intimacies in nonhuman animals in the first place, and (2) a critical approach to what this might demand in terms of challenging or transforming the ways in which humans engage in relationships of harm with nonhuman animals. This acknowledgement and critical reflection is a way to highlight the implications of emotional intimacy produced and experienced between members of other species. But it also requires a more inclusively multispecies, less anthropocentric understanding of emotion and intimacy.

Indeed, the risk of attributing what are seen as human characteristics to nonhuman lifeforms is often a primary objection to the consideration of animal emotion as a legitimate focus for research. Common concerns about anthropomorphism range from perspectives reflecting ideas about human exceptionalism (in which humans are seen as exceptional and completely unique in experiencing emotions and intellect) to concerns about not representing animals' experiences adequately by projecting human ideas onto what is being observed. Yet using human experiences need not mean disregarding the experience of nonhuman animals. Bekoff (2000, p.867) defines anthropomorphism as 'using human terms to explain animals' emotions or feelings [which enables] humans [to] make other animals' worlds accessible to themselves.' This accessibility is important as a mode of understanding nonhuman experiences, and Bekoff (p.867) reminds us that 'anthropomorphic language does not have to discount the animal's point of view'.

Understanding the point of view of other species involves a certain level of anthropomorphism and careful, indeed critical, reflection on feminist questions of how to represent authentically the perspective of another, especially when more usual methods (interviews or focus groups, for instance) are less available to the researcher. Being able to talk to, or read the words of another human being (even through a translator, or in translation), in order to try to understand another's perspective is not a method easily available in multispecies research. Instead, researchers must rely on observation, witnessing, bodily encounters and measurements, or interviews with human caretakers: all methods that multispecies ethnographers engage to gather knowledge about nonhuman lifeworlds. I argue, though, that a recognition of animal emotion as a form of nonhuman intimacy informs practical research considerations, opening space for particular forms of intimacy as a research practice.

Gruen (2015, p.24) argues '[t]hat we experience the world from a human perspective doesn't mean that we can't work to see things from the perspectives of nonhumans, and . . . empathy is a skill that helps us in doing this'. While research across species poses particular kinds of challenges, it also produces new possibilities for how feminist researchers and multispecies ethnographers might think about the role of intimacy, or empathetic relationships, in formulating new modes of knowledge-making. If, for instance, multispecies encounters enable building knowledge about other species through empathetic understanding and response, this might offer new insights into how intimacy is centred as a practice in feminist research. With this in mind, I turn to a second way of understanding intimacy: intimacy as empathetic research practice.

Intimacy as research practice

That other animals have complex emotional experiences of the world, and that these are impacted often and intensively by human actions, are insights that inform intimate research practice. But how might researchers develop this way of seeing? And how might researchers do so when so much of academic scholarship is dedicated both to discounting the role of emotion in scholarly research, and to reinforcing anthropocentric notions of human exceptionalism in terms of whose lives count as lives and whose emotional inner worlds are legitimate? I suggest that Gruen's (2015) framework of entangled empathy is a way of defining intimacy in feminist research practice in order to centre intimacy itself, as well as nonhuman lifeworlds, in feminist scholarship.

As a mode of defining intimacy in research practice, entangled empathy is articulated by Gruen (2015, p.3) as 'a type of caring perception focused on attending to another's experience of wellbeing. An experiential process involving a blend of emotion and cognition in which we recognize we are in relationships with others and are called upon to be responsive and responsible in these relationships by attending to another's needs, interests, desires, vulnerabilities, hopes and sensitivities.' I argue that engaging in this kind of empathy as caring perception as a research practice brings a level of intimacy into research that allows for greater attention to the lived experiences of those whom feminist researchers and multispecies ethnographers study.

As I have outlined above, an acknowledgement of animal emotion – manifest as a kind of intimacy between members of other species – is an important step in developing this kind of empathy as intimate research practice. But it is just that – a step; indeed, as Gruen (pp.51–52) writes:

> I think of empathy as a process. Although the process may not be linear, we can think of the various parts of the process as going something like this: The wellbeing of another grabs the empathizer's attention; then the empathizer reflectively imagines himself in the position of the other; and then he makes a judgment about how the conditions that the other finds herself in contribute to her state of mind or wellbeing. The empathizer will then carefully assess the situation and figure out what information is pertinent to empathize effectively with the being in question.

This sort of empathy doesn't separate emotion and cognition and will tend to lead to action because what draws our attention in the first place is another's experiential wellbeing. Once our perception starts the process, we will want to pay critical attention to the broader conditions that impact the wellbeing or flourishing of those with whom we are empathizing. This requires us to attend to things we might not have otherwise. Empathy of this sort requires gaining perspective and usually motivates the empathizer to act ethically.

I would like to suggest that incorporating this kind of empathy into research practice can generate more nuanced insights about the wellbeing and experiences of others (whether they are human or nonhuman). In the case of the cow and the calf, I was prompted to an empathetic response by the intimacy and trauma that was readily visible in their attachment to and separation from each other. In Gruen's formulation, this encounter grabbed my attention. I then tried to imagine myself in their position, letting myself try to feel what that experience might be like. Next, I tried to take a step back and look at the structural conditions that produced these embodied particularities for the cow and calf – conditions of domestication, commodification, and use – and then, I refocused on their physical and emotional states. I considered then, and in subsequent reflection on this encounter, what I might be missing.

This attention meant that I likely missed other things in the moments surrounding this encounter – focused so intently as I was on the cow and the calf themselves. For instance, I was not as attentive to how the other human spectators and buyers were responding to what was happening in the ring. I was not as attentive to the next few animals who passed through the ring, focused as I was on listening to the calls between the cow and calf that echoed forward from the back of the auction yard as they tried to communicate from their now-separate pens.

The systematic nature of the animals moving through the auction yard also posed an ethnographic challenge in terms of what I might be missing. I didn't see the cow and calf interact before arriving at the auction, nor did I see them, or know what happened to them, after they left the auction yard with their respective buyers. This is a problem related to the fractured and alienated lives of animals in commodity production, which is exacerbated by the geography of food industry practices. For instance, the segmentation of cows on dairy farms and bulls on semen-producing farms, and the removal of calves from cows shortly after birth enact routine forms of separation in animals' social networks and segmentation in the commodity circuit. Auction yards, in particular, operate on a spatial logic that severs these intimate bonds. The economic efficiency of the auction yard also renders the intimate worlds of the animals an abstraction, focusing instead on their reproductive and productive capabilities for commodity production and not the emotional interior effects of this thorough and routine commodification.

Thus, a space like the auction yard (much like other spaces of commodification in the dairy industry) poses a problem in terms of developing intimate research with cows used for dairy because of their thorough conceptualization as commodities and as living property. As I said above, at times, their physical or emotional pain was easily visible to me (as an observer attentive to their embodied responses). But the kinds of intimacies that are developed and known over time were difficult to access in ethnographic research where access to these animals' lives was almost always fleeting. And so the resulting intimate accounts of animals in my research were contingent, partial and incomplete. Sometimes I was witnessing just a moment in the lives of these animals as they passed through the auction pen. How to learn something meaningful about the animal in that moment requires prior knowledge of what animals go through in dairy production, such as extensive research on the process of dairy production, the gendered dimensions of the appropriation of animal bodies, and the process of sending spent animals to slaughter. This research and knowledge helped to fill out these short vignettes in a way that connected the animals' lives and experiences, and their emotional bonds with each other, to the broader economic logics governing animal bodies in the dairy industry more generally. And it also meant that the individual story – the moment in which the cow and calf were separated – became a lens through which to understand their suffering, as well as a way to perhaps understand the plight of other singular beings labouring and dying for food production in a sort of composite – albeit, incomplete – picture.

This incompleteness is, of course, part of any ethnographic account. Visweswaran (1994, p.1) writes, '[e]thnography, like fiction, no matter its pretense to present a self-contained narrative or cultural whole, remains incomplete and detached from the realms to which it points'. Ethnography always tells a particular story, from a particular perspective, representing particular kinds of entanglements between researcher and informant. Developing intimate research practice through empathy might be one way of creating a fuller picture, even (or maybe especially) in sites where glimpses into the lives of those who are being studied are fractured and fleeting. Of course, this kind of research, and the thinking and writing it prompts, takes time. Responding to and processing the emotional toll wrought by empathizing with those who are subjects of violence takes time. *Intimacy* takes time.

Can we rush intimacy? For slower research practices

My partner and I first met Saoirse, a small one-and-a-half-year-old beagle, in the anteroom outside of her kennel in the biomedical research laboratory where she was living. Her tail was completely tucked up under her in a canine expression of fear and submission, and she crouched low with her head down. Her forehead was wrinkled as she surveyed us warily from across the room. Her eyes were bloodshot, locked on us, watching. Her body was tense and quivering. The staff person who had introduced us left us in the room with her, and we sat there and waited patiently for her to come to us. She skirted the edges of the room and winced each time we moved. Finally, she crept up to me slowly and sniffed my hand. After a while of letting her explore and get closer to me, she let me pet her head and her ears. She had a tattoo in one ear with an alphanumeric identification combination and her belly was shaved from her recent spay surgery.

When we brought Saoirse home, she vomited in the car and she shook at every new noise, smell and sight, her body erupting into shivers. She had never been outdoors and the sensory experience of being outside the lab was a wholly overwhelming experience. She was like a ragdoll for days, sitting and shaking wherever we put her. It took weeks for her to relax in our home and bond with us. Going outside was almost too much for her at first. She would make it a few steps outside and then run back. Quickly, though, she learned to love running along through the grass, nose to the ground, taking in all the smells of the outdoors. Gradually, her personality came out and she became playful and active. She burrows under blankets to sleep and snores loudly, relaxing fully in her moments of rest. More than two years later, she is a different dog than the one we first met in the research lab. Occasionally, the traces of her time in the lab come out: when we go to the vet; when she encounters new people; when flashes of light streak across the ceiling from the early morning sun. But our shared life together has transformed the way she moves through and experiences the world around her. And it has transformed me, too.

Why intimacy as a research practice? Gruen (2015, p.25) explains in clear terms: 'harm . . . matters, but it does so in the context of a *particular* life. The abstract perspective allows us to overlook what is important from the other's point of view, and it also obscures the unique capacities that other animals possess and might be valued in themselves. Too often in this abstraction, we substitute our own judgements of what is beneficial for other animals for what may in fact promote their wellbeing.' Intimacy through empathy allows for attention to particularity. It allows for an attention to the *particular life* and its embodied experience of the structural conditions with which so many feminist researchers and multispecies ethnographers are concerned.

This level of attention was highlighted at first by the particular animals I encountered in the auction yards I visited, but adopting Saoirse while I was in the midst of my fieldwork on the dairy industry added another layer to thinking about intimacy and particularity. Meeting Saoirse, and the subsequent years sharing a home with her (and now two other beagles out of the biomedical research lab), has provided insights about intimacy in research that I did not expect: namely, that intimate research informed by entangled empathy is a way to access depth and moments of knowledge-making in fleeting or transient research sites. But more than that, it has highlighted the fact that this kind of intimacy is also more fully developed over time.

Empathizing with the cow and calf in the auction yard revealed a way to develop intimacy in my research practice in a place where intimacy was difficult to foster – it allowed for a focus on the *particular* life. The auction yard and its commodifying logic creates a level of fundamental abstraction from the singular animal, or the pair of animals in relationship with one another. An animal's singularity is typically visibly noted in the auction yard only insofar as it defines her commodity-producing potential. Empathy as a form of intimate research practice in this space can be a way to resist and better understand this logic of commodification, as well as the actual being it impacts.

I have not written an ethnography of my experience living in a relationship of mutual care with Saoirse, but if I did, it would involve a level of detail that was difficult – if not impossible – in the constraints of my ethnography of the dairy industry. Of course, we are always going to know someone (human or nonhuman) we live with much better than research subjects in a research site. But this experience illustrates the ways in which intimacy itself is an important mode of building knowledges about other species' lifeworlds. Our emotional states are intertwined – if I am anxious, she begins to show her own anxiety: forehead wrinkled, hyperactively running around the house and she whines. If she is anxious, and showing her anxiety in her embodiment, I can feel my own anxiety start to surface. I've grown to know her subtle bodily responses to the world around her in a way I didn't notice at first – how the way she sleeps reveals her level of relaxation, how minor differences in how she holds her tail betray her mood, or how the smell of her breath changes when she is afraid.

These details that we have learned about each other makes it possible for me to read external stimuli and the impacts of broader social and political economic relations more carefully and in a more nuanced way. In other words, seeing her in the lab, living with her in the aftermath of leaving the lab, and seeing her recuperation, I have been attentive to her place as a living being purposefully bred for, commodified and appropriated by, the biomedical research industry. As in the auction yard, I relied on empathy as a way to develop intimacy and knowledge about a nonhuman life. What Saoirse has highlighted especially is the exciting potential of intimacy developed over time in the research process. What I thought I knew about Saoirse after a few hours with her was soon eclipsed by what I thought I knew after a few days with her, then after a few weeks, followed by months, and then even now, by what I think I know about her after years together. What kind of intimacy as a form of knowledge-making will be possible after the course of a shared lifetime?

Intimacy as a form of research practice takes time. And when researchers are constrained in any number of ways (by limited access to the spaces that research subjects inhabit, by rushing to publish or produce to compete on the job market, and by time-consuming administrative and other under-recognized service to the university), time is hard to come by. This approach to thinking about intimacy – that it takes time, both as a research practice and as a form of sociality – aligns with the recent manifesto, 'For Slow Scholarship' (Mountz, et al., 2015). In it, Mountz et al. (2015, p.3) argue for a distinctly feminist approach to slow scholarship – scholarship that develops an ethic of care as it more intentionally takes 'time to think, write, read, research, analyse, edit, and collaborate'. This call for a more caring, slowed-down mode of scholarship is necessary for intimate research practice: for the time spent actively researching; for the time it takes to think and read more in response to what we've seen; for the time it takes to have a revelation that perhaps our approach was flawed and we need to go back to researching; and for the time it takes to write and process, perhaps before we even think about publishing. Taking time is one way we develop intimacy, and the challenges of multispecies ethnography emphasize the need for research which is attuned to intimacy and which takes the time to explore new ways of knowing, feeling and writing our scholarship.

By way of conclusion, these experiences of multispecies encounter with the cow and calf, and later with Saoirse, inform the need for research that is attuned to the intimate emotional worlds of nonhuman life. These experiences also show the importance of research that is attuned to empathy as a form of intimacy in feminist research and multispecies ethnography. And finally, the particularities of intimacy and how it is experienced and fostered as research practice prompt synergies with feminist geographers' call for slow scholarship. My hope is that the practice of intimacy through empathy might be fostered as a research praxis which simultaneously recognizes its incompleteness and contingencies *and* prompts transformative explorations of creative new ways of knowing how we are intimately entangled with others.

Works cited

Bekoff, Marc, 2000. Animal emotions: exploring passionate natures. *BioScience*, 50(10), pp.861–70.

Bekoff, Marc, 2007. *The emotional lives of animals: a leading scientist explores animal joy, sorrow and empathy – and why they matter*. Novato, CA: New World Library.

Gillespie, Kathryn, 2014. Sexualized violence and the gendered commodification of the animal body in Pacific Northwest US dairy production. *Gender, Place and Culture*, 21(10), pp.1321–37.

Gillespie, Kathryn, 2016a. Nonhuman animal resistance and the improprieties of live property. In: Irus Braverman, ed., *Animals, biopolitics, law: lively legalities*. New York: Routledge.

Gillespie, Kathryn, 2016b. Witnessing animal others: bearing witness, grief, and the political function of emotion. *Hypatia*, 31(3), pp. 573–88. doi:10.1111/hypa.12261.

Gruen, Lori, 2015. *Entangled empathy*. New York: Lantern Books.

Hatkoff, Amy, 2009. *The inner world of farm animals*. New York: Stewart, Tabori & Chang.

King, Barbara, 2013. *How animals grieve*. Chicago: University of Chicago Press.

Kirksey, Eben and Helmreich, Stefan, 2010. The emergence of multispecies ethnography. *Cultural Anthropology*, 25(4), pp.545–76.

Mountz, Alison, Bonds, Anne, Mansfield, Becky, Loyd, Jenna, Hyndman, Jennifer, Walton-Roberts, Margaret, Basu, Ranu, Whitson, Risa, Hawkins, Roberta, Hamilton, Trina and Curran, Winnifred, 2015. For slow scholarship: A feminist politics of resistance through collective action in the neoliberal university *ACME*, 14(4), pp.1–24. [e-journal] Available at http://ojs.unbc.ca/index.php/acme/article/view/1058 [Accessed 14 July 2016].

Pachirat, Timothy, 2011. *Every twelve seconds: industrialized slaughter and the politics of sight*. New Haven: Yale University Press.

Visweswaran, Kamala, 1994. *Fictions of feminist ethnography*. Minneapolis: University of Minnesota Press.

Part IV

Analytical methods as part of writing

15 Bearing witness to geographies of life and death

Intimate writing and violent geographies

Samuel Henkin

Introduction

I am witness,

> I am introduced to the story, or history, of a young man, though I will never meet this man his story crosses mine and becomes an unsung burden and a sense of him belongs to me. Standing in the snow of Sachsenhausen Concentration Camp, beyond the depths of any humanity, a pink triangle is forcefully sewn onto his shirt. Gay. The pink triangle consumes his body, his mind and his spirit, an annihilating virus. Every fiber of his being is tortured, hope fades, and he is left to bear witness. *Arbeit macht frei*, welded into the gates of his prison. Freedom? There is no freedom for him. I imagine the moments before death set upon him, the pain and suffering silenced. The life and love deprived from him parts into the world, his lover's voice whispering his name and he is once more clutched in the arms of his lover, an invisible compassion to light the way. (Henkin, 2010)

I believe that intimate writing can be useful in making sense of extreme violence. I want to speak of spaces and experiences of extreme violence – to cry out – and yet an overwhelming incomprehensibility suppresses my desire – not necessarily knowing how to speak of extreme violence. How does one elucidate experiences wherein the thresholds between life and death are violently rendered indistinguishable? In what ways can one intimately engage the violent erasures of subjectivity, humanity and life itself? As a researcher engaged in intimate understandings of extreme violence using, intimate writing holds the potential to speak of atrocities where bodies and self(s) are un-made through pain, torture, suffering and death (Scarry, 1985).

In contemplation of the questions above, I explore intimate writing within the context of researching temporally-bounded spaces of extreme violence where my autoethnographic knowledge and the historical narrative of Heinz Heger, a gay Holocaust survivor intersect. Intimate writing holds the potential to render the emotional milieu of experiences, understood through memory, of extreme violence visible and transmittable. I conceive of intimate writing as an act through which to understand the interconnected relationships between various geographical imaginations of the local, the global and the intimate that are embodied and lived.

Heinz Heger is the pen name of Josef Kohout, whose memoir 'The Men with the Pink Triangles' details his experiences and the experiences of gay men in the concentration camp system. As witness, Heger delivers his testimony of the extreme violence of the Holocaust through memory. Memory is a political practice that is both transitory and transformative and emphasizes socio-spatial experience and agency as a manifestation of an incontestable reality (Scarry, 1985). Memoirs – serving as personal, intimate memory vessels – are imbued with a multitude of emotional responses, relations and dimensions of one's lived experiences often understood as separate from ourselves (Griffin, 1993). Yet, through our own sentience the phenomenon of connectedness is unmistakably present.

Using intimate writing, both as a source, Heinz's memoir, and a methodology, an approach to analysing information, facilitates the teasing out of embodied emotional connectedness between Heinz Heger and myself. Through a reading of Heger's memoir I pull out the embodied emotional whereby my own intimate, subjective, and situated knowledge is readable alongside the material realities of Heger's narrative. A connectedness across time and space is communicated and constituted through such intimate situated knowledge, known as a sense of living connection (Hirsch, 2008). The materiality of the human experience is bounded within emotionally located moments through which these sensed connections are realized. Geographies that are emotionally located exist within the material spatial relations of everyday life and recognize the inevitable embodiment of geographical processes and geographical relationships (Dowler and Sharp, 2001, p.169; Tyner and Henkin, 2015).

I employ intimate writing as a way to express and understand my sense of living connection to Heinz Heger. Using intimate writing, my sense of living connection with Heger is conceptualized through postmemory, or the transmission of his memories to me in affective, embodied and spatial forms (Hirsch, 2008). His lived everyday experiences of violence, torture and death at Sachsenhausen Concentration Camp, a space of extreme violence that I came to experience sixty years later under extraordinarily different contexts, engendered an embodied and emotional transmission of memory through a common spatial sentience, though temporally displaced (Henkin, 2014). In closing, I discuss the possibilities of using intimate writing to read intimate writing through notions of postmemory and bearing witness.

A postmemory

In the era of 'Memorial Mania' (Doss, 2012), it is easy to see how obsession with memory contributes to building specific landscapes. The consequences of this obsession are in effect 'transmuted into history, or into myth' (Hoffman, 2011, p.xv). Memory and writing are inexorably intertwined in shaping history, particularly violent histories. The transmissions of violent memories through intimate writing forwards tangible emotional embodiments within shared spaces of extreme violence, in other words a sense of living connection.

Concerned about maintaining the sense of living connection, Hirsch (2008) uses postmemory in relation to the violent histories of the Holocaust as a way to understand 'the relationship of second generation to powerful, often traumatic, experiences that precede their births but that were nevertheless transmitted to them so deeply as to seem to constitute memories in their own right' (p.103). Applying postmemory beyond that of the second generation could be an effective way to channel memories of experience and knowledge introgenerationally. Postmemory as both a theoretical lens and a methodological tool is sensitive to the spaces and experiences of the everyday that encompasses material and affective interactions between people, places and how our realities are constructed and shaped (Shotter, 1993). The transmission of memory is evaluated by its affective, intimate and subjective power (Hirsch, 2008). Postmemory serves as part of the broader inter-relations of human experience and existence and our sense of living connection to others that can be read through intimate writing.

Postmemory understands the distinctions between the emotions, the embodiments and the places of extreme violence. For individuals who actively experienced extreme violence the thresholds differentiating each are indistinguishable. Heger's experiences were normalized by violence, pain and the overwhelming continuous confrontation of his mortality. In the camp there were moments in which life was lived explicitly through violence, torture, pain and death. Pain permeated through mundane and banal details of the everyday and buttressed the destruction of world and self experienced spatially by the contraction of one's universe to mind and body (Scarry, 1985; Anderson and Smith, 2001; Bondi, 2005; Tyner and Henkin, 2015). In the camp, which served as the Nazi regime's most violent spatial apex, dehumanization became both *literal* – destruction of bodies – and *figurative* – destruction of subjective qualities. Yet these violent erasures of humanity and subjectivities themselves serve as a means through which memory can be transmitted (Hirsch, 2008). Out of the perversion of a victim's violent total subjugation into the falsified assertion of Nazi power grew the perversity for resistance whereby spaces of the intimate – the body, emotion and mind – remain consciously sentient. Intimate writing expresses this knowledge and confronts the realities of this unimaginable occurrence, disrupting the very foundations of the camp (Felman and Laub, 1992). Intimate writing enables a resistive thread of narrative and memory that can be and has been circulating in the culture of our post-Holocaust generation for decades (Felman and Laub, 1992).

In January, 1940, Heinz Heger, marked with a pink triangle, a technique of categorization used to identify gay men, entered Sachsenhausen Concentration Camp through the front gatehouse (Heger, 1980). *Arbeit macht frei* was etched into the irongate, a material fixity of the threshold between humanity and the universe of the camp, between life and death (Sofsky, 1999; Hirsch, 2012). The violent and disciplined socialization process of entering the camp lasted six days, but the mental and physical deprivation and destruction inflicted by the Nazi regime continued; 'the intention was not just to kill us off immediately, but rather to torture us to death by a combination of terror and brutality, hunger and bitter toil' (Heger, 1980, p.44). Heger's sexuality marked him for death and served to situate violence of the unbearable directed against his body and self (Plant, 1986; Hirsch, 2012). Heger was liberated in 1945 after five years of experiencing the horrors of the concentration camp system. But for Heger and thousands of other pink-triangle survivors, liberation was not fulfilled in the fullest sense as persecution and violence only continued in a sphere of heteronormative and homophobic oppression (Heger, 1980; Seel, 1995; Spurlin, 2009). After 27 years of silence and denial bounded within a culture of impunity surrounding gay Holocaust survivors, Heinz Heger transcribed his memories as a testament to the thousands whose stories and memories could not be passed on, 'May they never be forgotten, these multitudes of dead, our anonymous, immortal martyrs' (Heger, 1980, p.118).

Over 60 years later, I, self-imagined as unmarked, entered Sachsenhausen Concentration Camp through the front gatehouse. *Arbeit macht frei* is etched into the irongate, a material fixity in the memorialized landscape of what was once a threshold between life and death. It was in barrack No. 39, which houses the permanent exhibition *The 'Everyday Life' of prisoners in Sachsenhausen Concentration Camp 1936-1945*, that I, with Heinz Heger's memoir in hand, intended to liberate his lived everyday experiences. I had hoped to ignite inspiration and understanding for the (LGBTQ) communities gay men are members of today where lives are continually disciplined, regulated and transformed by various geographies of oppressions (Henkin, 2014). Yet, this unfolding of events was not what transpired at Sachsenhausen Concentration Camp. Instead I perceived his memories – his intimate lived experiences – in *my own body*, with the power of situated and spatial awareness provided by my immediate spatial presence (Felman and Laub, 1992). Memories are anchored in time and place, and resonate through my body. The memories' legacies are traced in the memorialized landscape whereby the material connection between the past and present becomes the evidential force through which the physical presence within shared spaces exposes bodily or sensed memory and a sense of living connection (Assman, 2006; Hirsch, 2012).

Heger's intimate renderings carefully 'grafted indelibly into my own life story' (Young, 1997, cited in Hirsch, 2008, p.97). Within the spaces of Sachsenhausen Concentration Camp, I became aware of the ways in which my body is marked: a transformative moment whereby my initial coming out materialized as Heger's past life abetted in reshaping my future (Hirsch, 2008). I became aware of my emotional embodied connectedness to his intimate writing, as I walked the same ground. In turn, I embarked on my own intimate writing piece, resolving the exploitive, violent and oppressive regimes of knowledge in my life with affirmations of belonging, 'For I as a gay man, marked by society by my own sexuality, bear within me a pink triangle' (Henkin, 2014, p.115).

Bearing witness and intimate writing

Intimate writing situates the ambiguities of disembodied accounts of violence within a geographical framework that seeks to understand complex intimate social relations (Pain and Staeheli, 2014). Extreme violence is more than just momentary aberrations of pain; it is sustained by organized practices, intelligibilities, rationalities and techniques. Often times its untranslatability and silence, as well as cultures of impunity, serve to obscure. Intimate writing affords an opportunity to unshackle extreme violence from various situated positions and bring it back from oblivion. Heger's intimate writing salvaged historical experiences from over half a century of amnesia and suppression and in doing so bears witness to its presence. Bearing witness permits the extrication of victims from extreme violence and the veils of untranslatability and silences, as well a culture of impunity through reconditioning their humanity. It 'connects us, and obligates us, to each other' (Oliver, 2001, p.20), and, by our very existence, we show, testify and ascribe meanings to world(s). To be witness is to become a bearer of history, an 'embodiment of memory (*un homme-mémoire*), attesting to the past and to the continuing presence of the past' (Wieviorka, 2006, p.88). Heger assumed the role of witness through his intimate writing. So do I.

Moving beyond simple narration, to bearing witness, bringing history into the present, is a commitment to others and an acknowledgment of the realization of varied experiences and occurrences (Felman and Laub, 1992). Expressly, bearing witness is not circumscribed by the production of explicit testimony concerning the condition of life; rather, it is a phenomenon that serves as a reminder that 'nothing faced by the human – however degrading and dehumanising – is condemned to oblivion without trace while there is witness' (Lechte and Newman, 2012, p.526). Heger witnessed the atrocities of the camp and his memoir in turn bears witness. His acts prevent the erasure of humanity pursued within the Nazi regime's geographical imagination.

Agamben (2002) situates impossibility as confronting the logical contradiction of the limits of what is possible. Impossibility exists beyond the limits of reasoning, paralleled in the seemingly unreasonable and unconscionable experiences of the concentration camp as a space of extreme violence (Lechte and Newman, 2012). The act of bearing witness to experiences of what reaches the limits of the possible embodies a resolution of not necessarily knowing but actively acquiring an understanding of what *not* knowing means: 'whoever assumes the charge of bearing witness in their name knows that he or she must bear witness in the name of the impossibility of bearing witness' (Agamben, 2002, p.34; Felman and Laub, 1992). The Holocaust is understood as unimaginable and unspeakable in nature. Intimate writing has the potential to illustrate how language can be used to bear witness in the face of the unspeakable. In bearing witness through intimate writing, it can be recognized that memory and its transmission is a powerful medium in survival and resistance that transgresses individual experiences and bounded spaces of extreme violence.

Bearing witness through intimate writing can be an important basis for human subjectivity, whereby through our very own existence(s) we engage in the creation of knowledge (Felman and Laub, 1992; Griffin, 1993, p.164). At Sachsenhausen Concentration Camp I connected emotionally to the embodied presence of a witness to the limit of the possible – the unbearable. What I have come to realize is that while bearing witness to Heger's past, in an attempt to grasp how spaces of extreme violence come to be, the strands of my understanding coalesced, forming a string, one which grows thicker with each exposure to the similar acts, events and practices in the institutions of violence that have come to life in the present.

For me, through a more personal understanding of postmemory I came to embrace a connective sociospatial politics across generations. Owing to intimate writing, I have been able to boldly fill the gaps and silences characteristic of spaces and experiences of extreme violence as a way to bear witness. Intimate writing of experiences, embodiments and spaces of extreme violence can lead to better understandings of what has happened in the past and serves as a reminder that violence and resistance are experienced in our present.

Conclusion

My practice of intimate writing has engaged violence, pain and suffering in embodied and memorialized forms to expose the subtle shallows and vast depths of violence. It has unveiled for me the ways in which the thresholds between life and death are difficult to understand on my own (Sylvester, 2013). By uniting theoretical understandings and methodological techniques within intimate writings of postmemory and bearing witness, the legacies of violence are no longer sanitized and censored within normalized logics, discourses and imaginations. Logical, ethical and moral questions are fundamentally entrenched in understanding the ways in which intimate, memorialized and embodied knowledge is embedded within relational social structures and processes.

I never met Heinz Heger. Yet our material paths crossed over 60 years apart. I embraced the affective power of his narrative across a series of emotionally located transformative moments and spaces. The process of bearing witness through intimate writing rendered connections visible. As practices, when used together, both have the potential to be an important emotional release at a deeply personal level that constructs senses of living connections across time and space. 'The darkness no longer lingers as my senses regain meaning, someone inside me emerges and is revealed, distinguishable from my own secret suffering. Silence no longer impairs me. This is who I am' (Henkin, 2010).

Works cited

Agamben, Giorgio, 2002. *Remnants of Auschwitz: the witness and the archive.* Translated from German by Daniel Heller-Roazen. New York: Zone Books.

Anderson, Kay and Smith, Susan J., 2001. Editorial: emotional geographies. *Transactions of the Institute of British Geographers*, 26(1), pp.7–10.

Assman, Aleida, 2006. *Der lange Schatten der Vergangenheit Erinnerungskultur und Geschichtspolitik.* Munich: Beck.

Bondi, Liz, 2005. Making connections and thinking through emotions: between geography and psychotherapy. *Transactions of the Institute of British Geographers*, 30(4), pp.433–48.

Doss, Erika, 2012. *Memorial mania: public feeling in America.* Chicago: University of Chicago Press.

Dowler, Lorraine and Sharp, Joanne, 2001. A feminist geopolitics? *Space and Polity*, 5, pp.156–76.

Felman, Shoshana and Laub, Dori, 1992. *Testimony: crises of witnessing in literature, psychoanalysis, and history.* New York: Routledge.

Griffin, Susan, 1993. *A chorus of stones: the private life of war.* New York: Anchor Books.

Heger, Heinz, 1980. *The men with the pink triangle: the true life-and-death story of homosexuals in the Nazi death camps.* Translated from German by David Fernbach. New York: Alyson Books.

Henkin, Samuel, 2010. *Coming out.* [letter]. (Personal communication, 31 January 2010).

Henkin, Samuel, 2014. *From camps to closets: geographies of oppression.* MA. Kent State University.

Hirsch, Marianne, 2008. The generation of postmemory. *Poetics Today*, 29(1), pp.103–28.

Hirsch, Marianne, 2012. *The generation of postmemory: writing and visual culture after the Holocaust.* New York: Columbia University Press.

Hoffman, Eva, 2011. *After such knowledge: memory, history, and the legacy of the Holocaust.* New York: Public Affairs.

Lechte, John and Newman, Saul, 2012. Agamben, Arendt and human rights: bearing witness to the human. *European Journal of Social Theory*, 15(4), pp.522–36.

Oliver, Kelly, 2001. *Witnessing: beyond recognition.* Minneapolis: University of Minnesota Press.

Pain, Rachel and Staeheli, Lynn, 2014. Introduction: intimacy-geopolitics and violence. *Area*, 46(2), pp.344–60.

Plant, Richard, 1986. *The pink triangle: the Nazi war against homosexuals.* New York: Henry Holt and Company.

Scarry, Elaine, 1985. *The body in pain: the making and unmaking of the world.* Oxford: Oxford University Press.

Seel, Pierre, 1994. *Liberation was for others: memoirs of a gay survivor of the Nazi Holocaust.* Translated from French by Joachim Neugroschel. New York City: Da Capo Press.

Shotter, John, 1993. *Conversational realities: constructing life through language.* New York: Sage.

Sofsky, Wolfgang, 1999. *The order of terror: the concentration camp.* Princeton: Princeton University Press.

Spurlin, William J., ed., 2009. *Lost intimacies: gender, sexuality and culture.* New York: Peter Lang.

Sylvester, Christine, 2013. *War as experience: contributions from international relations and feminist analysis.* London: Routledge.

Tyner, James A. and Henkin, Samuel, 2015. Feminist geopolitics, everyday death, and the emotional geographies of Dang Thuy Tram. *Gender, Place and Culture*, 22(2), pp.288–303.

Wieviorka, Annette, 2006. *The era of the witness.* Translated from German by Jared Stark. Ithaca, NY: Cornell University Press.

Young, James E., 1997. Between history and memory: the uncanny voices of historian and survivor. *History and Memory*, 9(1/2), pp.47–58.

16 Becoming fieldnotes

Ebru Ustundag

Introduction

In 2010, I became involved with a local charity organization that sought to provide a safe drop-in space for former and current female sex workers in the basement of a local church in St. Catharines, Ontario. Since then, that drop-in has opened its doors every Thursday between 9pm and 12am. The organization provides healthy meals, hygiene products and clothing for female street-level sex workers. Throughout the night, drop-in volunteers, of which I am one, do outreach on the street, distributing hygiene packages (including condoms, socks, underwear and hard candy) as well as snacks and drinks around Queenston Street in downtown St. Catharines. One of the effects of volunteering at the drop-in was that I became part of the everyday lives of these women. In a short period of time, I established working and personal relationships with various social service providers in Niagara Region including Positive Living Niagara, Segue Clinic, Niagara Region Health Public Health, Quest Community Health Center, YWCA of St. Catharines and John Howard Society of Niagara. I also worked with other volunteers on grant proposals for the drop-in to secure financial support from Niagara Community Foundation as well as United Way of St. Catharines and District. In addition to these formal processes related to the drop-in, I have accompanied the women to medical, legal and social service provision appointments, including detox centres and court dates. As part of my research and my scholarly activism, I have also played a role in co-organizing International Day to End Violence Against Sex Workers events. I have given several presentations on sex work in St. Catharines for various local organizations including local churches and women's organizations.

At some point in my navigations of social spaces and identities, I started taking notes – all hand-written – of various moments, events and emotions that I witnessed or was part of. These notes made their way into what I called *my field notes*, reflecting some of my emotional struggles around understandings of everyday practices of outdoor sex workers. In the writings, I had been hoping to reflect some of my struggles around understanding the complexities of relationships and dynamics that were unfolding around me about the everyday lives of a group of marginalized people in a relatively conservative environment with limited access to social and financial resources.

At some other point, I realized that I was writing in two languages, Turkish (my first language) and English (my academic language). I was surprised! I have been in English-speaking Canada for seventeen years and my academic training had always been in English. Yet while revisiting and organizing my field notes, I was intrigued by how I was switching from one language to another while taking these detailed notes of my thoughts, observations and encounters. Once I thought about it, I expected that the content of my nights as a volunteer would be more academic information and would be inscribed in English. In contrast, my feelings of despair, anger, anxiety, helplessness and resentment were more personal and would be written in Turkish. But this was not the case. In fact, I couldn't find a common thread to explain the pattern of my transitions. I was a bit unsettled to see that my assumptions about my sense of my own identities were wrong. The field notes called into question my own narrative practice.

In trying to understand my own reflective research practice, I looked into scholarship on linguistics and psychology about bilingualism. Pavlenko (2006; 2014) underlines that various languages are not identifiable and countable entities. She maintains that languages are not easily identifiable, discrete or countable entities. She argues that 'there is no such thing as the bilingual mind: bilinguals vary greatly in linguistic repertoires, histories, and abilities, and the bilingual mind appears here as an umbrella term to refer to a variety of speakers including multilinguals' (Pavlenko, 2014, p.ix). Pavlenko's contributions to the conceptualizations of how emotion and emotion-laden words are represented and processed in bi-and multilingual mental lexicons deeply resonated with how I was beginning to understand my own use of English and Turkish. Her intriguing work made me curious as to how positionality, reflexivity and subjectivity might have different connotations and means of expression in English and Turkish. I found myself asking questions regarding feminist understandings of spatial becomings and subjectivities, and in turn began mapping out the disruptions, tensions and contradictions in my field notes. I turned to the research process as a contested one and questioned understandings of field notes as so-called brute data. The messi-ness of research processes is well-documented (Moss, 2002; England, 2006) and I decided that I wanted to come to terms with some of that mess, the mess I found in my field notes.

In this chapter, I reflect on my engagement in critical research practices, espe-cially in how they relate to field notes. I first explore the relationship between and field notes and subjectivities. I then discuss how the framework I offer to explain the emergence of subjectivities in the field helped me to understand my own field notes in relation to disruptions in narrations, expectations and representations. I end the chapter by providing some themes that build upon the idea of what I define as becoming fieldnotes.

Field notes and subjectivity

Feminist, critical race and indigenous scholars' contributions to understanding power relations in relation to self, identity, subjectivity and knowledge production have been shaping contemporary theoretical and methodological discussions in geographical thought (Moss and Falconer Al-Hindi, 2008). Alongside the deconstruction of what research itself is, a significant intervention in these discussions has been the destabilization of what is understood as the field (e.g. Katz, 1994). Throughout my several years of participant observation, social justice activism and community organizing, I have navigated various positionalities in disparate spaces composed of multiple micro-geographies of street-level sex workers in St. Catharines. Various academic studies unpack the complexity of sex work from various standpoints, including liberal and abolitionist views (e.g., Augustin, 2005; Ditmore, Levy and Willman, 2010). This complexity is also evident in reports by sex worker-based organizations across Canada, such as POWER: Prostitutes of Ottawa/Gatineau Work, Educate and Resist; Maggies: Toronto Sex Workers Action Project; Stella: Making Space for Working Women; Pivot Legal Society; and Pace: Providing Alternatives, Counselling and Education Society. Benoit and Shaver (2006) note that sex work research is generally quite broad in its implications and covers various issues around social and labour rights, addiction and health care, and social exclusion and marginalization. Yet there is a growing literature on the role of space in the constitution of sexual identities and sex work (e.g. Hubbard, Matthews and Scoular, 2008; Hubbard, et al., 2008). Though most sex work is now sold off the streets (Hubbard, et al., 2008), understandings of micro-geographies of outdoor sex workers without moralistic, victimizing and whorephobic frameworks are still limited.

Like many other academics and social justice activists (Nagar and Geiger, 2007; Wright, 2010), I have been interested in exploring the relations of micro-level everyday life to larger power relations (Staeheli and Nagel, 2006; Pratt, 2012) and emerging new female subjectivities (Oksala, 2013). I am most interested in detailing the political possibilities arising out of the *messiness* and *complexities* of the research, including navigating diverse emotions, positionalities and spatio-temporalities when the everyday life of different groups becomes entangled. Highlighting how field notes can be read as expressions of the multiple subject positionings we take up as researchers can facilitate how feminist approaches to research in geography enable a re-thinking of field notes as a political practice of resistance. As academic disciplines are self-governing and operate within neoliberal discourses, researchers are expected to have clear objectives and anticipated results of all research projects. This *modus operandi* not only shapes various stages of research practices, especially the design and circulation of findings, but also governs researchers' care of the self (Rogers, 2012). Under these governing practices, I argue that as part of a critical praxis-oriented approach to research, inscribing detailed field notes in two different languages provides a space of resistance in the constitution of subjectivities. Indeed, analysing field notes can be an entry point into identifying and understanding researchers' multiple yet contradictory subjectivities.

Critical praxis-oriented research as a political project has become even more crucial for scholars who are interested in unpacking the complexities of everyday life of marginalized groups. The invisibility of material and political struggles of the social reproduction of everyday life of women has been an important field of inquiry of social citizenship. My field notes in both English and Turkish reflect the discrepancies in our social and health systems and how it has been challenging for groups of marginalized people to access the services they need. Of course this is not peculiar to St. Catharines. Within neoliberal and neoconservative discursive regimes, vulnerable populations face various barriers to accessing various social services in Canada. However, it is significant here to underline the fact that it has been even more challenging for vulnerable groups in relatively smaller cities where resources and services are significantly limited.

A fundamental part of ethonographic research, field notes are generally conceptualized as brute data. They are seen as connecting researchers to their subjects. Thinking field notes beyond data construction and collection and beyond mere descriptions of our research subjects is vital to a critical praxis-oriented research. They are more complicated and more nuanced than merely unproblematic links; they are expressions of the multiple subject positionings that one takes up as a researcher. The intimate process of writing field notes reflects not only the messiness and complexities of research processes in the field on the fly, but also how one navigates the tensions and contradictions that emerge from those processes. Anthropologists stress how field notes are an important communication tool with oneself, as well as symbols of professional identities. Jackson (1990) shows how an anthropologist's engagement with field notes varies greatly – from data to contextualization. She also points out how field notes and processes of intimate writing have been important tools for researchers to constitute their identities as scholars. Writing field notes also complicates the assumptions one might make about identities. What I propose here is that the processes through which writings in notebooks, on scraps of paper and in word processing files are influx themselves, and that this process, *becoming fieldnotes*, is a site of resistance.

Field notes as resistance

Most of my field notes reflect issues around privilege and reflexivity. Like the anthropologists that were mentioned above, my field notes reflect the tensions, paradoxes and contradictions of multiple positionalities that I have been navigating. After five hours at the drop-in on Thursday nights, including a couple of walks around the track (the area where sex workers and clients meet) to distribute water, snacks and hygiene packages to women (and to whomever else might need them), I have had a really tough time adjusting to going home to my bed with clean sheets and the ongoing realization that I am fortunate enough to have food in my fridge, when people I deeply care about and love don't have much. It becomes even more challenging when on Friday mornings, I get on the bus to go to the university where I take up the positioning of a self-governing neoliberal academic. My everyday materiality in my home wrenches me from the subject positioning I had been inhabiting all evening. Engaging in teaching, writing and service work at the university jerks me into another nomadic space. The issue here is not so much that I value one over another (for each has its own set of contradictions that I experience as both positive and negative), it is more that it makes my head spin having to make such dramatic shifts in such short periods of time. Yet being nomadic allows me to think through and move across interconnectedness (Braidotti, 1994). Of course, such tensions in the life of a scholar/activist and activist/scholar are not unique or at all new. Rather, for me, the way that these emotions are transcribed in field notes reveals what kind of a researcher, scholar and an ally I want to be and become.

Another recurring theme in my field notes is my frustration regarding my struggles with my own (in)abilities to deal with the various positionalities I have been navigating. For example, my personal and political engagement in these relationships became even more complicated during the fall of 2012 and winter of 2013 when I was on sabbatical. Being away from teaching and day-to-day university service enabled me to be more available on a regular basis for phone calls, face-to-face chats and appointments. This made me even more connected to what was happening on the streets and what impact the changes have on street-level sex workers in St. Catharines. For example, in March 2013, we started seeing large abscesses on women's bodies as a result of levamisole, a lethal chemical used to cut crack and cocaine. Given the social stigma and stereotyping in various service provision centres, it was challenging for street-level sex workers to access health services. My interaction with street-level sex workers extended beyond my weekly work at the drop-in and around the track. I was also in touch with some women who had been off the streets for a while and were not interested in attending the drop-in any more. These women generally tried to be away from the track and crack houses in order to stay safe and clean, though they frequently admitted how much they missed their connections and friendships in these spaces. When women decide to stay away, and if they wish to stay in touch with me, I meet with them – sometimes just to have coffee, maybe to bring hygiene products from the drop-in, and sometimes to refer them to social and physical/mental services in the area in consultation with doctors I am in contact with. Though I am now convinced that the best way to stay sober and off the streets, if that is what one desires, is to be away from St. Catharines for good. Of course, this is easier said than done. Most of the women miss their friendships and connections that they have built over the years with others on the streets, in crack houses and in prison or jail. However, if they come back after being clean for months or even years, they unfortunately relapse over a relatively short period of time. Even though I know that revisiting social networks is not the only reason for a relapse, social relations and the desire to belong to a community play an important role in women returning to the streets for sex work to support their addiction. These intimate encounters and experiences reflect not only the contested nature of sex work, but also how sex workers' everyday lives are enabled and disabled via various scales of power dynamics. Field notes emphasize the fact that nomadic subjects are always in transition and don't belong anywhere.

My field notes offer valuable insights regarding to socio-spatial and spatio-temporal constructions of various subjectivities. The field in this research includes physical spaces like the drop-in, hospitals (including emergency rooms), pharmacies, courthouse, women's detox, cemeteries, funeral homes, crack houses, restaurants, alleyways and downtown St. Catharines' streets and parks, as well as digital spaces such as Facebook. Navigating multiple research sites can be challenging for researchers (Mullings, 1999; Fisher, 2014). While engaged in this research, it has been challenging to move beyond dualistic constraints of myself and to situate my subjectivity within a particular theoretical configuration. When I think about temporalities of street-level sex work, I often find myself questioning linear and conventional notions of time, as the women's everyday practices and social reproduction are organized in various socio-temporalities. A common misconception about street-level sex work is that it is only a nighttime activity. In St. Catharines, one of the busiest times is around 3:30pm, when the shift changes at the local General Motors plant. The so-called afternoon rush brings a flow of clients to the track. The night crowd is active until the beginnings of daylight. Temporalities of the week might also be read differently. Thursday, the night of the drop-in, is essential for those who depend on the groceries and hygiene products provided there. The drop-ins are important spaces for belonging and social comfort for some, but not for those who might have to encounter other women with whom they have personal conflicts. The flow of women who frequent the drop-in can also change at the end of the month, when government checks are issued. Variation in the seasons affect living on the street and visit to the drop-in. As expected, winters are difficult, but summers are too. With people being outside more and being excessively drunk, women waiting on the corner to be picked up are subjected to violence from random people driving by the track. What doesn't change with the season is how women and other drug users use alleyways and streets to inject drugs under unsanitary conditions. Recording everyday lives of sex workers accompanied by the deconstruction of various stereotypes about sex work and sex workers has been a significant part of my field notes.

One of the things that I keep repeating in my field notes has been to pay close attention to what is not seen and/or heard on the streets. As Braidotti (1994, p.16) argues, 'writing is not only a process of translation but also successive adaptations to different cultural realities'. In this regard, writing as a process is unsettling and disconcerting, and always in flux. While my initial field notes include detailed description of certain events and my reactions to them, as years passed by I began purposefully paying more attention to writing about the things that I left out earlier. Braidotti (1994, p.17) calls this the nomad who knows how to read invisible maps: 'the map is invisible or, rather, it is available to those who have been trained to read invisible ink signs'. This reflexive process has become more of a challenge when some of the sex workers disappear for a period of time, for various reasons. Some come back to St. Catharines, some don't. The complexities of entanglement of the micro-geographies of the everyday in my life has made me question essentialized notions of identity that inform positionality and reflexivity. These intense representations of multiple subject positionings via detailed recordings of everyday encounters of various social groups have demonstrated the nomadic nature of the subject in the field notes. The field notes merged – becoming fieldnotes: ones that were no longer brute linkages or static entities, but were fragments of emergent subjects.

Becoming fieldnotes across spaces not only reflects moving beyond dualistic constraints of the nomadic subject but also emphasizes the potentiality of interconnectedness and affinities between and among various subjectivities.

Further reflections

The embodied act and process of writing that I have been engaged in on a regular basis facilitates a witnessing of various articulations of my various subject positionings. Writing about the conflicts, despairs and dilemmas of complex relationships made me realize that my field notes are not, indeed cannot be depicted as, only a record of data collection in one's research. As Braidotti (1994, p.4) iterates, 'whereas identity is a bounded, ego-indexed habit of fixing and capitalizing on one's selfhood, subjectivity is a socially mediated process of relations with multiple relations with multiple others, and with multilayered social structures'. Intimate writing via field notes is a significant example of the constitution of subjectivities in and through various spatialities as it contests self-imposed identities (e.g. activist, scholar activist, friend, ally). Working with and through various emotions including inadequacy, disappointment, guilt, confusion, despair and hope provides spaces of vulnerability, resistance and healing. In this process research itself becomes healing, while field notes become resistance. By rejecting essentialized understandings of the self, Braidotti's nomadism offers novel ways of thinking about contemporary subjectivity as well as interconnectedness and affinities between various social groups across spaces. When field notes become fieldnotes there is a potential and possibility for political and ethical transformation via 'blurring boundaries without burning bridges' (Braidotti, 1994, p.4).

Acknowledgements

I would like to express my heartfelt gratitude to Pamela Moss and Courtney Donovan, for their invitation to write this chapter, and for Pamela Moss's generous editorial support throughout its various stages.

Works cited

Augustin, Laura, 2005. The cultural study of commercial sex. *Sexualities*, 8(5), pp.618–31.
Benoit, Cecilia and Shaver, Frances M., 2006. Critical issues and new directions in sex work research. *The Canadian Review of Sociology and Anthropology*, 43(3), pp.243–51.
Braidotti, Rosi, 1994. *Nomadic subjects: embodiment and sexual difference in contemporary feminist theory*. New York: Columbia University Press.
Cupples, Julie, 2002. The field as landscape of desire: sex and sexuality in geographical fieldwork. *Area*, 34(4), pp.382–90.
Ditmore, Melissa Hope, Levy, Antonia and Willman, Alys, eds., 2010. *Sex work matters: exploring money, power and intimacy in the sex industry*. London: Zed Books.

England, Kim, 2006. Producing feminist geographies: theory, methodologies and research strategies. In: Stuart Aitkin and Gill Valentine, eds., *Approaches to human geography*. Thousand Oaks, CA: Sage. pp.286–97.

Fisher, Karen, 2014. Positionality, subjectivity, and race in transnational and transcultural geographical research. *Gender, Place and Culture*, 22(4), pp.456–73.

Hubbard, Phil, Matthews, Roger and Scoular, Jane, 2008. Regulating sex work in the EU: prostitute women and new spaces of exclusion. *Gender, Place and Culture*, 15(2), pp.137–52.

Hubbard, Phil, Matthews, Roger, Scoular, Jane and Augustin, Laura, 2008. Away from prying eyes? The urban geographies of adult entertainment. *Progress in Human Geography*, 32(3), pp.363–81.

Jackson, Jean, E., 1990. 'I am a fieldnote': fieldnotes as a symbol of professional identity. In: Roger Sanjek, ed., *Fieldnotes: the makings of anthropology*, pp.3–33. Ithaca: Cornell University Press.

Katz, Cindi, 1994. Playing the field: questions of fieldwork in geography. *Professional Geographer*, 46(1), pp.67–72.

Moss, Pamela and Falconer Al-Hindi, Karen, eds., 2008. *Feminisms in geography: rethinking space, place and knowledges*. Lanham, MD: Rowman and Littlefield.

Moss, Pamela, ed., 2002. *Feminist geography in practice: research and methods*. Oxford: Blackwell.

Mullings, Beverly, 1999. Insider or outsider, both or neither: some dilemmas of interviewing in a cross-cultural setting. *Geoforum*, 30(4), pp.337–50.

Nagar, Richa and Geiger, Susan, 2007. Reflexivity and positionality in feminist fieldwork revisited. In: Adam Tickell, Eric Sheppard, Jamie Peck and Trevor Barnes, eds., *Politics and practice in economic geography*. London: Sage. pp.267–78.

Oksala, Johanna, 2013. Feminism and neoliberal governmentality. *Foucault Studies*, [online] 16, pp.32–53. Available at: http://rauli.cbs.dk/index.php/foucault-studies/article/view/4116 [Accessed 16 November 2016].

Pavlenko, Aneta, ed., 2006. *Bilungual minds: emotional experience, expression and representation*. Clevedon, UK: Multilingual Matters Ltd.

Pavlenko, Aneta, 2014. *Bilungual mind*. Cambridge: Cambridge University Press.

Pratt, Geraldine, 2012. *Families apart: migrant mothers and conflict of labor and love*. Minneapolis: University of Minnesota Press.

Rogers, Dallas, 2012. Research, practice and space between: care of the self within neoliberalized institutions. *Cultural Studies* ⬄ *Critical Methodologies*, 12(3), pp.242–54.

Smith, Andrea, 2006. Heteropatriarchy and the three pillars of white supremacy. In: Incite! Women of color against violence, ed., *Color of Violence: INCITE! Women of Color against violence*. Cambridge, MA: South End Press, pp.66–73.

Staeheli, Lynn A. and Nagel, Caroline, 2006. Topographies of home and citizenship: Arab-American activists in the Unites States. *Environment and Planning A*, 38(9), pp.1599–614.

Sterry, David Henry and Martin, Jr., R.J., eds., 2009. *Hos, hookers, call girls and rent boys: professionals writing on life, love, money, and sex*. Berkeley: Soft Skull Press.

Wright, Melissa, 2008. Gender and geography: knowledge and activism across the intimately global. *Progress in Human Geography*, 33(3), pp.379–86.

Wright, Melissa, 2010. Geography and gender: feminism and a feeling of justice. *Progress in Human Geography*, 34(6), pp.818–27.

17 Hiding in the garden

Autoethnography and intimate spaces

Kathryn Besio

Prologue

I haven't seen anything written about lurking as a research method, probably because it's suggestive of some creepy stalker. My dictionary defines it as 'to remain hidden so as to wait in ambush for someone or something' and that resembles what I do, walking around the neighbourhoods of Hilo, looking at households' vegetable gardens, hoping that someone will come out, and engage me in conversation about their yard. I don't want to intrude uninvited into their private spaces, so I wait for an invitation. It's an ethnographic ambush: I observe quietly, passively hoping for participation. Unfortunately, lurking achieves observation but works less well for participation. I see what people grow in their yards, whether they have vegetable beds or just fruit trees scattered about, and walking around the public edges of people's homes convinces me of the ubiquity of food plants in Hilo yards. It's a landscape reading exercise that I do during my physical exercise and it's nice to combine my interests in fitness and geography.

Strangely I've found lurking a more successful method in my local museum's collection, the Lyman Mission House and Museum in Hilo. I didn't know exactly what I would find, if I would find anything at all there, and my goals weren't well-defined. I thought a look at whatever they had might yield something. Searching the museum's collection implies intention and direction, and because I didn't know what they had in the archive, I'm hesitant to call what I did a search. Like my walks through the neighbourhoods of Hilo, I wandered through scrapbooks and folios, hoping that something would jump out at me from the hodgepodge of documents therein. And something did ambush me when I came upon a nineteenth-century photograph entitled, *Scene in Hassinger Garden* (see Fig. 17.1). That photograph began a search for information about it and a person who may have been lurking, although I don't know. I can't say why he's there, but now that I've seen him, it's hard to look away.

My finding in the archive made me think more about my relationship to living and researching in Hawai'i, pushing me to reflect on the ways I hide my research in plain sight. I walk around the edges of household gardens for reasons both embarrassing and anxious. It embarrasses me that I am not more direct in recruiting research subjects, feeling the need for them to come to me rather

Figure 17.1 Scene in Hassinger Garden, photographer unknown, probably J.A.
        Hassinger, ca. 1890–1899

Source: Courtesy of the Lyman Museum.

than just knocking on their door. But the anxious reason why I don't knock is
particular to living in Hilo, Hawai'i. Although I have lived in the Pacific for over
twenty years, I remain wary of the micro-confrontations of ethnicity, class and
privilege that are part of postcolonial daily life, interactions that are in many ways
easier to deal with if they're not considered research. As a non-Native Hawaiian
researcher, I often feel uncomfortable about my research interests. Shouldn't I be
doing something more activist? But what? The past stalks the present in Hawai'i
as elsewhere, but we don't always take time to reflect on it, although lurking gives
me time to ponder further.

Introduction

The photograph, *Scene in Hassinger Garden*, is the catalyst for this chapter. It
made me reflect on my research methods and through that reflection, I saw connections between my day-to-day existence in Hilo and how I do research here.
In this chapter, I rely upon autoethnography (Besio and Butz, 2004; Besio, 2005;

Butz and Besio, 2004; 2009) to examine the significance of this reflection. I use autoethnography in multiple ways: first, as my own reflective autoethnography and second, as subaltern autoethnography in the photograph. Subaltern autoethnography is when marginalized subjects insert themselves into the discursive spaces of the dominant, such as texts and photographs, and in idioms that may not be their own but to which they have access. This latter form of autoethnography draws from various researchers (most notably Pratt, 1992). I have not applied insights from autoethnography to my previous writing and research on Hawai'i, at least not directly. But in this chapter, autoethnography helps me to think more clearly about the linkages between intimate (homes and gardens) and public spaces (streets and towns), the past and present, and research and not-research.

I didn't go looking for the *Scene in the Hassinger Garden* and, like the gardens I walk by in Hilo, it *became* my data. The photograph began a search for unknown and in all likelihood unknowable subjects: the enigmatic man in the bananas and the women in the garden setting. Homes and gardens often go unmentioned in historical sources, being outside the ken of historical geography (Morin and Berg, 1999; Blunt and Rose, 1994). I also saw the photo as an example of a subaltern autoethnography, that is, a photograph in which marginalized subjects used an idiom that is not their own but for their own representational purposes. It seemed to me that the women and man in the bananas expressed something about their histories in Hawai'i in a photograph they most likely didn't compose. The photograph's colonial tableau engages viewers in a narrative that illustrates Hawai'i's ethnic diversity, although the image may appear in the first instance as a racialized binary of white and non-white. Like other examples of subaltern ethnography that researchers may use to try to understand (Pratt 1992), but *unlike* those where the researcher may be present through participant-observation (Butz, 2001; Besio, 2005), the subjects in the photograph are in a paradoxical space: they are both at the centre and at the margin. They are marks in the historical record, making my assumptions about what their presences, expressions and intentions mean an ontological challenge, although not one without epistemological merit. That is, how does what I imagine from this photograph, an intentionally- critically- and contextually-informed imagining of the garden scene, follow from what I know about Hawai'i? Further, how does what I learn from the photograph then influence my analysis of Hilo's contemporary gardens and intimate spaces like them? These questions come together in the photograph that I happened to have found tucked away in a folder of a small local archive.

My social context influences my intellectual *and* emotional engagement with my research sources, just as my research informs my social life, and the contemporary gardens of Hilo intersect with images of gardens in the archives. What was it about the photograph that so grabbed my attention? Why couldn't I look away? Reflective and subaltern autoethnography helps me *make sense* of the links between past and present and here and there. In this chapter, I draw upon my previous work in autoethnography, linking it to new directions in autoethnography and reflective writing such as DeLysers's (2015) archival autoethnography, and the reflective writings of de Leeuw (2012). In combining autoethnography and the reflective writing analysis by de Leeuw most specifically, I tease out some of the difficulties in researching intimate spaces like home gardens in postcolonial settings like Hilo. Before I turn to my reading of the photograph, a bit of groundwork in the research of autoethnography, DeLyser's archival autoethnography and de Leeuw's reflective writing, is helpful.

Archives and autoethnography

The prologue to this chapter is a reflective narrative, highlighting some of my conflicted feelings about research and living in Hawai'i, and it is a type of autoethnographic writing that examines my own practices and experiences, written in the first person (Ellis and Bochner, 2000). The *auto* in this use of autoethnography refers to the ethnographizing of the researcher, and situates my research interests in the photograph, contextualizing the photo as part of a larger project on home gardens. It also conveys something about my personality and positionality that often goes unsaid in academic writing, but that is important to how I do research. As noted, geographers may turn to more personalized writing such as this not only to locate themselves within their projects, but also because reflexivity offers an important analytical tool (see Valentine, 1998; Cook, 2001 for excellent examples).

DeLyser's (2015) work on archival autoethnography brings autoethnography into dialogue with historical research. I find particularly interesting her claim that historical geographers often treat archives as merely 'extractive resources': namely, as places that researchers mine and dig around for information, ignoring the constructed nature of the archive itself (DeLyser, 2015, p.210). In this metaphor, archives are cordoned off as separable veins of data where an intentional drawdown takes place, rather than a space where we are always looking (lurking?) for information. One might add that field research is often portrayed in a similar way, as a location separate from daily lives where one goes to retrieve information, although feminist researchers have long disputed this division. Like the Hilo neighbourhoods where I walk, the Lyman Museum archive is a part of the field where I hoped to find historical material, although I really didn't know what I might find there. While the photograph is now a part of my data, it was sheer luck I found it, largely due to the archivist pulling the folder for my viewing. Maybe this counts as extraction, but it was certainly not predictable that anything would come from this field of information. Perhaps it really wasn't chance exactly, but there certainly was a fortuitous element involved.

Unlike DeLyser, my autoethnographic interest in archival research is not in how I created an archive, but in how the data from the archive influenced my research. From her archival ethnography, I gained a greater appreciation for contingency and acknowledgment that what becomes data is as much due to the researcher's choice, or perhaps in my case, the archivist's choice, as the item itself. DeLyser traces the creation of her *own* collection of Ramona tourist kitsch, illustrating that collections are 'contingently produced' (DeLyser, 2015, p.210) based upon the availability of items, the interests of collectors and the sorts of items that are deemed collectable at different times. She reflects on her own collecting habits to show that her postcards and collectables became data and not just tourist kitsch once they became part of *her* collection, showing another way that researchers' engagements with archival sources go beyond extraction and into construction.

Analogous to DeLyser's collection, the scrapbooks and photographs at the Lyman Museum *became* data because they crossed my path, becoming useful to me. What took me to the museum was that I had read that in order to provide foods for their families, Hilo's North American missionaries had planted home gardens, learning from Native Hawaiians about plants that would thrive in the tropical climate unfamiliar to the New England missionaries. There's a much longer transcultural narrative that underpins this cultivation story that I can't go into here, but the short version is that American missionaries left records about their home lives, not just in published journals available in libraries, but in the form of scrapbooks, unpublished letters, and so on, that are buried in the museum. These items are not on display. They hide, perhaps lurking in the back offices of the museum.

The Lyman Museum is named for David and Sarah Lyman, Hawai'i Island's first American missionaries. Like other of Hawai'i's nineteenth-century missionaries from the American Board of Commissioners for Foreign Missions (ABCFM), the Lymans came from the northeastern US. The ABCFM missionaries had varying experience with missionary work (Kashay, 2007; Schulz, 2014), but most had never left the US until they came to Hawai'i, and this was true of the Lymans, who came to Hawai'i soon after they married. The Lymans stayed in Hawai'i until their deaths. Two of their seven children stayed in Hawai'i, and like their parents and other missionary offspring, Hawai'i became their home. The items housed in the Lyman Museum reflect an entangled narrative of missionaries and their others in transcultural Hawai'i.

Based upon my conversations with the museum's archivists, I learnt that the archive is an evolving rather than a cohesive collection of Lyman memorabilia. The collection grows as new acquisitions arrive, and many of the items are placed into folders that contain an array of photos, newspaper clippings and heirlooms from donors unrelated to the Lymans as well as unknown donors. The items have been archived by numerous archivists, who do not have good records of who donated what items or when they were donated. This is not to cast aspersions on the archivists who do their best as part-time employees working in an underfunded non-profit facility but to say that finding an item's provenance may be difficult. According to the museum's current archivist, the majority of the museum's holdings come from Hawai'i Island donors, although donations come from residents from other Hawaiian islands and the continental states. Donors find items that they send to the museum, because they don't want to keep them, believing that items related to Hawai'i Island should find a home there. Although the archive's primary purpose is to house materials related to the Lyman family and their descendants, there are many items in the archive whose provenance is unknown.

Thus it's a mistake to assume that the items in the Lyman Museum archive reflect a unified missionary vision, although the collection is dedicated to housing documents related to Hilo's missionary and plantation era residents. I highlight the uncertain provenance of the items in the archive, because in name, the museum ostensibly focuses on the missionary colonists in Hawai'i, although that focus initially misled me. I assumed that the items in the museum were exclusively of missionary descent but they are not, although it's fair to say that the collection's focus is on the materials of non-native descendants, reflecting settlers' and missionaries' perspectives. Additionally, many of Hawai'i's English language historical sources are in the records, journals and publications of missionaries, capitalists, bureaucrats and their descendants in small archives like the Lyman Museum, across the state. My use of materials in English limits the historical data available to me as a non-Hawaiian speaker, which then influences my representations of Hawai'i's colonial past because it's one based on writings by non-native colonists. As de Leeuw (2012, p.3) maintains, 'working through questions about British Columbia's colonial past, frequently in efforts to more fully understand the province's neocolonial present, means often working with records and archives reflecting the dominance of European settler colonialism.' With respect to using non-indigenous records this is as true for researchers working in English in Hawai'i as it is for those in British Columbia.

However, the Lyman Museum and similar missionary-related institutions in Hawai'i may contain holdings from a diverse group of settler imperialists, and facile distinctions between Native Hawaiian and non-Native Hawaiian become apparent. What may be less obvious because of the missionary/Native Hawaiian binary that frames Hawai'i's history is that the American missionaries and the non-missionary foreign populations often had interests at odds with one another, even as they were sometimes at odds with Hawai'i's indigenous Native Hawaiian population (Daws, 1968). For example, the conflict over temperance between missionaries and sailors is well-known; another example is that missionary families worshiped at churches separate from both Native Hawaiians *and* non-missionaries, like sailors and merchants. The diversity in Hawai'i's non-native history is less apparent today, likely due to the fact that once the Hawaiian monarchy was overthrown in 1865, settlers' and imperialists' interests coalesced and collapsed into a racialized dualism of non-native white (*haole*) and Native Hawaiian interests. The simplistic binary of *haole* and Native Hawaiian dominates much of the historical and contemporary political discussion about Hawai'i, with much of the diversity of Hawai'i's non-native, that is, mixed ethnicity plantation descendent populations, both white and non-white lurking in the background. In de Leeuw's (2012, p.275) attempt to understand British Columbia's diverse colonial relations, she uses reflective writing to offer a 'committed, impassioned and emotive response to the archival record.' Her empathetic and personalized response to her archival subject, Alice Ravenhill, helps her depict colonialism's narrative as nuanced and diverse, much as I am trying to do to in Hawai'i. Reflexive writing, such as in my prologue, is in de Leeuw's words an 'emotive response' to the historical record. Yet the photograph allowed me to bring my emotive autoethnographic response into conversation with subaltern autoethnography, another form of autoethnographic representation.

Postcolonial researchers often look to archival sources for instances in which those who are excluded from the written record may make their presences known in instances of resistance to colonial control. These instances may take the form of subaltern appropriations of colonial idiom that could be called subaltern autoethnography (Butz and MacDonald, 2001). Subaltern autoethnography is a form of writing-from-below and uses the language or practices of the dominant group to re-present a marginalized perspective. I don't want to overstate that in applying an autoethnographic sensibility to the Hassinger photograph I can ever know the definitive story about the making of the image or how it came to be in the Lyman Museum archive, or that I will ever have the ability to say much about the subjects in the garden and their representational intentions. I don't know why the photographer may have chosen to print this photograph that contains the women or figure on the margin. Yet it is important to be attuned to the possibility of subaltern autoethnography as I read the materials from this archive. In as much as the *Scene in the Hassinger Garden* may be an instance of subaltern autoethnography, I think my autoethnographic approach makes room for multiple narratives to emerge: the subjects in the photo and my own narrative of the present. It is to the photograph that I now turn.

The scene

Within the Lyman Museum archive's folders and scrapbooks there is a tension between unity and haphazardness, public records and intimate documents. I found myself looking through folders with titles like, 'Geography, Hawai'i, Hilo, originally 1900' or 'Domestic Lives, Homes'. There are newspaper photographs of events and places, and personal photographs of homes, often of interiors with well-laid tables surrounded by stunning amounts of bric-a-brac. My first thought upon seeing the latter was how did they keep this stuff from getting mouldy in Hilo's humid, wet climate? My second thought, more germane to research, was that in these folders is a pictorial discourse of so-called civilized people displaying their civilized lives to themselves and future generations. For example, in one scrapbook, pages show striking juxtapositions, offering an example of what one scrapbooker deemed important to memorialize, raising questions about the person's worldview. In it, the scrapbooker juxtaposed a photo of the dining room and an aerial photo on one page and a Native Hawaiian *mahu* (transgender person) on the next.

What I saw in the collection fascinated me, and as I looked through scrapbook after scrapbook and folder after folder, my main impression was that in these scrapbooks, like the neighbourhoods I wandered through, I was still peering in from the outside, without interpretation or explanation by those I was looking in upon. I was still lurking. Unlike looking at a scrapbook while sitting side by side with its maker, I couldn't ask why they included the images they did; at least wandering through Hilo, I could, if inclined, knock on the door and ask. I was left to imagine: what were they thinking? Why were the images organized the way they were? Yet it was a folder entitled 'Domestic Lives, Homes E–H' that stopped me in my tracks (Figures 17.1, 17.2 and 17.3).

Handwritten across the top of the photo is the title: *Scene in Hassinger Garden*. Three adults stare out from its centre, surrounded by banana plants. There is another adult, near the right edge of the photograph, apparently a native man, who looks different from those in the centre. At first glance, the photograph is a startling colonial tableau. The group at the centre appear to be of American or European descent, looking directly towards the camera, engaging the viewer forwardly, particularly the younger woman seated to the left, whose penetrating gaze is partially obscured by shadows and foliage. A bearded male, holding what I think is a light metering device, reclines in the centre of the composition between the women. Behind him is a seated woman, shaded by the banana fronds. Hiding in the fronds is a child wearing striped socks, and whether the child is male or female it is hard to tell. She or he is well-hidden, a ghosted double-exposure; it appears that the child moved while being photographed. I didn't see the child until months later when I received a digital copy of the photograph. The sepia-tinted image – I'm still not sure what process was used for the photograph – is over-exposed at the top, adding to its ethereal quality.

Figure 17.2 Close-up from *Scene in Hassinger Garden*

Source: Courtesy of Lyman Museum.

Figure 17.3 Close-up of male figure from *Scene in Hassinger Garden*
Source: Courtesy of Lyman Museum.

Immediately, the photograph reminded me of a prudish, tropical version of Edouard Manet's *Déjeuner sur l'herbe*, although other people subsequently viewing the image have told me the photograph is much more reminiscent to them of Paul Gaugin's paintings of Tahiti. Perhaps it is the tropical garden setting and the way that the younger woman in the Hassinger photo looks at the viewer, or perhaps it's the man in the bananas who calls Tahiti to mind. I recalled the *Déjeuner* because much of what I recall about that particular painting is about gender relations, which is where my fixation on Manet rather than Gaugin stems. Yet the Hassinger photo women are unlike those in the *Déjeuner*. In the photograph, the women sit erect in their chairs, hands folded at their knees, and there is nothing louche about them, like the charges levelled at the women in Manet's *Déjeuner*. The *Déjeuner* originally hung in the Salon de Refusés amongst other pieces but was rejected by the 1863 Salon because of its subject matter, as well as the style in which it was painted. In the painting, one of the women is half-naked, picnicking with fully dressed males, and the other partially clothed splashing in the creek behind. I recalled that art historians read *Déjeuner* as contradictory commentary on gender equality: the naked woman stares at her viewers unashamed, as if daring them to look at her, while simultaneously, because she is nude, she is read as objectified by the masculine gaze. I wondered if the photographer had been aware of the painting in composing his photograph, because it has a similar awkward intimacy. The reclined, sprawled male body at the centre contrasts with the women's direct gaze and erect poses. His almost casual pose at the centre alludes to a form of masculine power now referred to as manspreading. That is, his pose reflects an entitlement of masculine bodies to spread out and take up space, while female bodies as feminine subjects sit holding their bodies inwards, embodying a subordinate position (Young, 1990). In the photograph, even as the women sit above him on their chairs, they remain beneath him in status, physically holding their bodies tightly together.

The ambiguity in the Hassinger photo women's pose recalls the Manet painting because of the contrast between their assured gaze and the white male's claim to space at the centre of the photograph. The photo differs from other posed garden photographs I had looked at in the Lyman collection, such as the one of Reverend David Lyman and his wife, Sarah, sitting in front of their home (Fig. 17.4). The upright, posed figures in the Lyman photo appear to convey values consistent with the missionaries' Calvinism. In contrast, the Hassinger photo, while not improper in comparison to the *Déjeuner*, may have suggested moral laxity to viewers of the late 1800s because of the reclining male at the centre and the direct gaze of the women, even as the image startled me.

Of course, it is not solely the nineteenth-century masculinist composition that makes the photograph so compelling to look at. In fact, the gender narrative at the centre may be secondary to the racialization of the image, which relies on the contrast between the figures in the centre and at the margin. What makes the photograph riveting is the inclusion of the non-white figure who is in but not wholly

Figure 17.4 Lyman Mission House
Source: Courtesy of Lyman Museum

in the photograph. He looks at the camera, peering out from behind a banana plant, the image capturing his hovering disembodied head. Viewers have to look carefully to see his shoulder and leg near the trunk of the banana plant. He does not appear be a family member because he looks different from the group in the centre, although he appears relaxed, even a bit playful, almost like a contemporary photo-bomber as he may have stepped into and out of the image. The man in the banana plants looks at the viewer, as if to say, 'I'm here, too, don't look away.' My initial impression of the photograph was that the man in the bananas was intended as a native presence. He wasn't an oblique reference as a native figure in a photograph juxtaposed to missionary objects such as, dining room tables, but a part of the photograph's composition. He demands the viewer's attention, albeit slyly, and his presence makes this gendered imperialist tableau unforgettable.

I left the museum, obsessing about the image, and as soon as I got home, looked at it again. My second-guessing began. The archivist had let me photograph the items I had looked at in the folders, so that when it came time to order better quality ones and request reproduction rights, I would have a record. The more I looked at the photo, the less the man in the bananas looked ethnically Hawaiian and the more he looked to be from somewhere else in the Pacific. I thought perhaps Micronesia or Samoa, but as I looked longer, other places emerged, Melanesia or Fiji. At that realization, the photograph became much more complicated, and in many ways more reflective of transcultural and multi-ethnic narratives of planta-tion Hawai'i. However, in multi-ethnic nineteenth century Hawai'i, there were very few Melanesians, Fijians, Samoans or Micronesians, and the larger migra-tions of those Pacific Islanders did not occur until the twentieth century. My assumption about his ethnicity was not the only one that I had incorrect, and in my initial reading I should have been more attentive to the context of Hawai'i's hybrid and pluralistic race relations that are rarely black and white, non-native/ Native Hawaiian or missionary/Native Hawaiian.

I have no way of knowing the intentions of the man in the banana plants or of the women in the photograph. Although from an autoethnographic perspective, I assume that he and the women may have used the idiom of photography to insert himself into a narrative not of his nor the women's own making and into the social space of the family grouped in the centre. Just as there is an ambiguity in the photograph's gender relations, the presence of the man in the banana plants cannot be read simply as a native man hiding and peering out, as a marker of tropi-cal authenticity or as imperialist composition of an apparently white family and a brown native. The photographer may have placed him in the photo as an indicator of Native Hawaiian presence, although he does not appear to be Hawaiian. Like the woman who looks directly at the camera, his forthright gaze disrupts any sim-ple reading of why he is there, lurking at the edges.

Conclusions from the garden

Based on the information that I have, the *Scene in Hassinger Garden* is most likely a family portrait taken by John A. Hassinger, the man in the centre. Finding information about Hassinger wasn't too difficult and his published obituary with his photograph confirmed his identity (Anonymous, 1902). However, there is scant information about him or his family elsewhere in the historical record, because he wasn't an important figure in Hawai'i's history. He arrived in Hawai'i as a purser on sailing ship and became a bureaucrat, initially serving under the Hawaiian monarchy and later under the territorial usurpers. He was never a missionary, and his obituary makes much of his activities as a Freemason (for more on the transcultural relations/history of Freemasonry in Hawai'i, see Karpiel, 2000). Hassinger was, though, an amateur photographer and a founding member of the Hawai'i Camera Club, which is why I think he staged and printed this family portrait. Additionally, and based upon other photos, I believe the women in the photograph are Hassinger's wife, Priscilla, and his daughter, Juanita, who modelled for him. I have not determined who the child is, perhaps one of Hassinger's other children. Of the man hiding in the bananas, I still know nothing of his identity, but I am still searching.

Contrary to my initial impression, and according to census and death records about Juanita Hassinger, she is listed as mixed race and as octoroon because her mother's ethnicity is Tahitian. Those who see Gaugin in the photo may see something of Hassinger's intention that his Tahitian wife and his *hapa* (the Hawaiian word for mixed-race) daughter stand out by posing them in a tropical garden. It could have been his intention to use the garden symbolically to link his wife and daughters to Tahiti, or even to link himself as a photographer to Gaugin. I have no evidence at this point to show that was his intention. Perhaps like Gaugin, he may have used the banana plants to exoticize and emphasize the women's difference, although that runs contrary to expectations about US race relations in the nineteenth century. Why would Hassinger want his family to stand out as non-white if they could pass as white and accrue its privileges? Adding to what I think I know about the family's ethnicity doesn't really challenge my initial readings of masculine superiority, although it reaffirms my understanding of Hawai'i as a place where binaries of race tend to breakdown into more hybrid categories.

Unfortunately, my research thus far has not led me any closer to finding out about the man in the banana plants. Was he, as I initially thought, inserting himself into the photograph's narrative, or did the photographer put him into the image to authenticate the family's Hawai'i narrative? That is, here is a native in the tropics in contrast to the family in the centre. Or was his blackness intended to whiten Hassinger's mixed-race daughter and his Tahitian wife? If that is the case, then the garden setting may have been chosen to heighten the man's otherness and thereby diminish Hassinger's wife and daughters' difference, as they sit upright in the centre of the image, looking a bit starched and stiff. Finally, why is the youngest member of the photograph almost completely hidden?

There are numerous evocative stories to tell about this photograph, and I find myself still lurking around the archive, both digital and material, in search of information about an image that seems to represent so much about a particular period of Hawai'i's history, but about which I can only imagine and say very little definitively. In spite of this looming absence of information, the photo has led me to a more reflective appreciation of Hawai'i's historical resources and the stories that emerge from them, particularly from intimate spaces like homes and household gardens. In all likelihood, the *Scene in Hassinger Garden* was a photograph for the family taken in their own backyard. It may not have been intended for a public audience, yet much can be gleaned from it.

If subaltern autoethnography is understood as text or performance by a subordinated subject, that is, a representation or self-narrative in an idiom not their own but to which they have access, then the Hassinger photograph evidences the presence of a wider range of non-white residents in Hawai'i than I previously assumed. Subaltern autoethnography 'has ontological implications in that it requires scholars to understand research participants as reflexive subjects whose self-narrations and indeed identities are constituted in relation to their own field that encompasses and entangles both parties' (Besio and Butz, 2009, 1668). When David and I wrote this statement, we were thinking about face-to-face ethnographic relations. But historical researchers are no less entangled with their subjects, and have fewer ways of finding information about their intentionality. For example, the appearance of the man in the banana plants could have been the intention of the photographer and how he composed the image. But unlike a painter or writer, the photographer did not necessarily control the gesture of the subjects during the photograph. Subjects may smile slyly or move or otherwise disrupt the moment of composition, reflecting something unintended by the photographer, which may be interpreted as autoethnographic gestures by those in the photograph. In printing the photograph, the photographer expresses his or her engagement with a range of his photographic subjects' self-narratives by tacitly approving the image. In this way, a family portrait expresses something more than a family's intimate image, and may have reached beyond the frame to the present.

Echoes of the past have a way of entwining with the present in our everyday lives and in our research practices. The Hassinger photograph grabbed me, forcing my attention in new directions, making me think about the significance of how I research, interpret and analyse the intimate spaces of people's lives, both in my own neighbourhood and in the archive. My reflexive autoethnographic writing provides that contextual detail about my own intimate research geographies, which became far more uncertain and contingent than I had realized at the outset. My prologue illustrates that walking through the neighbourhoods hasn't led me to many conversations, or any deeper into the social spaces of homes and gardens, although, perhaps ironically, the photograph allowed me into spaces I haven't yet accessed. The *Scene in Hassinger Garden* helped me connect my present practices to stories yet unheard from the places where I live work and live.

Epilogue

I continue to walk through Hilo's neighbourhoods, reading the landscapes of gardens. I have more questions than answers, not unlike my reading of the photograph. Lurking around the edges of home gardens is more than a metaphor of how I research Hawai'i as a space of unequal but shared experiences and the transcultural space described by scholars (Pratt, 1992; Butz & Besio, 2004). Lurking also describes my relationship to Hilo and to Hawai'i, a place where I, like others, may always feel a bit on the outside looking in. Sometimes I would prefer to have greater distance from my field. but I can't, or that I might live in a place with a less-contentious history, but I don't, although all places have their own contentious contexts. Like many research settings, intellectual and proximate distancing is unobtainable, despite ongoing mythologies of separateness. Living and researching in Hawai'i opens up opportunities for reflection alongside nervous anticipation. A project on home food gardens may seem an odd way to engage with Hawai'i's contemporary politics of colonialism and race. Yet it remains that much can be learnt about wider social relations from the embodied interactions in the homes, gardens and neighbourhoods, the spaces of daily life.

Works cited

Anonymous, 1902. John Adair Hassinger dies after long illness. *Commercial Advertiser*, 7 June, p.9.

Besio, Kathryn, 2005. Telling stories to hear autoethnography: researching women's lives in Northern Pakistan. *Gender Place and Culture*, 12(3), pp.317–32.

Besio, Kathryn and Butz, David, 2004. Autoethnography: a limited endorsement. *The Professional Geographer*, 56, pp.432–8.

Blunt, Alison and Rose, Gillan, eds., 1994. *Writing women and space: colonial and post-colonial geographies*. London: Guildford Press.

Butz, David, 2001. Autobiography, autoethnography and intersubjectivity: analyzing communication in northern pakistan. In: Pamela Moss, ed., *Placing Autobiography in Geogography*. Syracuse: Syracuse University Press. pp.149–66

Butz, David and Besio, Kathryn, 2004. The value of ethnography for field research in transcultural settings. *Professional Geographer*, 56(3), pp.350–60.

Butz, David and Besio, Kathryn, 2009. Autoethnography. *Geography Compass*, 3, pp.1660–74.

Butz, David and MacDonald, Kenneth, 2001. Serving Sahibs with pony and pen: the discursive uses of 'Native authenticity.' *Environment and Planning D: Society and Space*, 19, pp.179–201.

Cook, Ian, 2001. 'You want to be careful you don't end up like Ian. He's all over the place'. In: Pamela Moss, ed., *Placing autobiography in geography*. Syracuse: Syracuse University Press. pp.99–120

Daws, Gavin, 1968. *Shoal of time*. Honolulu: The University of Hawai'i Press.

de Leeuw, Sarah, 2012. Alice through the looking glass: emotion, personal connection, and reading colonial archives along the grain. *Journal of Historical Geography*, 38, pp.273–81.

DeLyser, Dydia, 2015. Collecting, kitsch and the intimate geographies of social memory: a story of archival autoethnography. *Transactions of the Institute of British Geographers*, 40, pp.209–22.

Ellis, Carolyn and Bochner, Arthur P., 2000. Autoethnography, personal narrative, reflexivity: researcher as subject. In: Norman Denzin and Yvonna S. Lincoln, eds., *The handbook of qualitative research*. Thousand Oaks, CA: Sage. pp.733–68.

Karpiel, Frank, 2000. Mystic ties of brotherhood: freemasonry, ritual, and Hawaiian Royalty in the nineteenth century. *Pacific Historical Review*, 69, pp.357–97.

Kashay, Jennifer Fish, 2007. Agents of imperialism: missionaries and merchants in early nineteenth-century Hawaii. *The New England Quarterly*, 80, pp.280–98.

Morin, Karen M. and Berg, Lawrence,1999. Emplacing current trends in feminist historical geography. *Gender, Place and Culture*, 6(4), pp.311–30.

Pratt, Mary Louise, 1992. *Imperial eyes: travel writing and transculturation*. London: Routledge.

Schulz, Joy, 2014. Birthing empire: economies of childrearing and the formation of American colonialism in Hawai'i, 1820–1848, *Diplomatic History*, 38, pp.895–925.

Stoler, Ann Laura, 2009. *Along the archival grain: epistemic anxieties and colonial common sense*. Princeton: Princeton University Press.

Valentine, Gill, 1998. 'Sticks and stones may break my bones': a personal geography of harassment, *Antipode*, 30, pp.305–32.

Young, Iris Marion, 1990. *Throwing like a girl and other essays in feminist philosophy and social theory*. Bloomington: Indiana University Press.

18 Death, dying and decision-making in an intensive care unit

Tracing micro-connections through auto-methods

Pamela Moss

Entering the realm of ICUs

As with any care responsibility, when things go well, everything runs smoothly. But when there is a glitch, things can unravel quickly.

For two weeks in the fall of 2009 she lay in an intensive care unit (ICU) drifting in and out of consciousness. Two quite serious processes were making her ill and two physicians from different branches of medicine were overseeing her medical needs. An infectious disease specialist, Dr Mikkelson, discovered that she had developed a gram-positive anaerobic infection on her hip prosthesis. A full replacement is the primary way to treat such infections. After a gut-wrenching 36 hours, I decided that no, she was too weak and would probably not survive the surgery. I opted for life-long antibiotics to keep the bacteria at bay. A lung specialist, Dr Viscaine, understood that the stroke had affected her brain stem, the site where breathing is regulated. Her sensors had been damaged, and she could only breathe properly when she was awake. The physician prescribed a bilevel positive airway pressure machine (BiPap) for 72 hours, and afterwards, she was to wear it only when she slept. A BiPap is like a portable ventilator that uses a mask over the face instead of a tube down the windpipe; it forces oxygen into the lungs and, by adjusting the air pressure, pulls carbon dioxide out.

Both doctors insisted that a family member be at her side during the 72-hour period in order to keep her in bed with the mask on. So I stayed, with only bathroom breaks when nurses were in the room. I slept when she slept. When she was awake, she cried or had visions, as she called them. I acted as a verbal restraint, telling her over and over again to quit taking the mask off. At the end of those three days, I went back to her apartment and slept uninterrupted for 30 hours.

A rationale for auto-methods

These opening paragraphs, a common form through which autobiographical writing manifests in feminist geography, do what they are supposed to do: set up a problematic around which to discuss a topic (see e.g. Valentine, 1998; Moss, 2001b; Longhurst, 2012). By problematic, I mean something akin to Smith's definition (2005). She argues that a particular experience acts as an entry point into a study of institutions. A researcher uses this experience to investigate how texts and discourses organize the banalities and mundaneness of everyday practices in order to access ruling relations. My problematic, however, neither acts as an entry point into the organization of work in an ICU, nor grants me access to ruling relations. Rather, I use problematic as an entry point into a discussion about how elements seemingly outside but still related to the social organization of an institution may affect the practices in the institution itself. In other words, this problematic permits me to discuss how my decisions and actions shifted the way the work in an ICU took place in particular moments.

As I write this, I feel torn. I want to write about something specific. I want to write about how the organization of work in an intensive care unit imposes excessive restrictions on the way that I as a family representative make decisions about death and dying on behalf of a loved one, or, in other words, how I approach the intimacies of caring, death and dying. I want to argue that these restrictions not only impose stringent parameters around the flow of information, but also leave room for ICU doctors and nurses to manipulate a desired outcome. I also want to write about something general. Something that shows how my experience is embedded in wider processes that are obscured from my view. I had become well-acquainted with the socio-legal regime of healthcare of a southern state in the United States (after Arrigio, 2002). I too came to understand ICUs as part of an apparatus of healthcare in a Foucauldian sense (Foucault, 1980a, p.194): 'a thoroughly heterogeneous ensemble consisting of discourse, institutions, architectural forms, regulatory decisions, laws, administrative measures, scientific statements, philosophical, moral and philanthropic propositions – in short, the said as much as the unsaid'. Thus I became interested in sorting through the connections between some of these diverse elements within healthcare systems; less in the sense of understanding the nature of the connection and the purpose of the apparatus itself (as in healthcare meets an urgent need of society) (p.195), and more in the sense of tracing how one specific connection gets made (Moss and Prince, 2014, p.31).

Tracings of micro-connections between, say, understandings of treating prosthetic infections and emotional attachments to loved ones can expose the glue that keeps the parts of an apparatus hanging together to meet the demands of society (such as a socio-legal regime). The same tracings can also show where connections have broken off, dissolved or become old-fashioned. These loose fittings permit the introduction of discursive and non-discursive elements that can alter the way in which an apparatus functions, or the way in which one negotiates the elements within that apparatus. Healthcare as an apparatus is embodied as a 'specific slice of matter', one that includes the discursive and the non-discursive as well as the said and the unsaid (Braidotti, 2013, p.35). Regulatory practices in medical treatment based on biomedical knowledge, such as the rules for recording vitals, writing prescriptions and giving orders, close off an embodied presence of the patient. They authorize a self-referential game of truth whereby the arbiters of what counts as true are the record-keepers themselves (Foucault, 1980b, p.131). Yet resistance is possible because truth is an effect of the practice of power (p.131). So while ICU patients are discursively present through objects in protocols and entries in file folders, their materialized presence is something open to ongoing negotiation, sometimes on a minute by minute basis, by everyone who is part of providing care (cf. discussion in Moss and Prince, 2014, p.187, where the weary warrior is materially present and discursively absent).

Auto-methods in feminist geography

Common understandings of methods in feminist geography using the self as a cornerstone around which to develop an analysis tend to place writing autobiographically as a way to expose truths in a life and to place writing autoethnographically as a way to position the researcher in the context of the field (Moss, 2001a; Besio, 2005; Butz and Besio, 2009). Most criticism methodologically holds that a focus on the self is navel-gazing, self-indulgent or serves only a therapeutic purpose for the author (Richardson, 2011). Critics in feminist geography sometimes consider stories that expose intimate moments in a life as contributing to a partial truth masquerading as transparent knowledge or an act that elides more politically pressing issues (see Kobayashi, 2009).

Yet auto-methods do have something to offer beyond the idiosyncratic. Even though they do not produce complete accounts of any one life or any one's life, through choices about what to include and what not to include, researchers using auto-methods can construct an analytical account, story or narrative that contributes to wider debates about numerous topics (Moss, 2001a; Sharkey, 2004; Bondi, 2014). In feminist geography, analysts often use autobiography to reveal spatialized aspects of some social phenomena, such as emotions, bodily limits or neoliberalizing institutions (e.g. Valentine, 1998; Longhurst, 2012; Moss, 2014). For me, writing autobiographically has been a way to sketch out some micro-connections among the diverse elements that make up various apparatuses I engage with daily (e.g. Moss, 2013). Thus, rather than a complete life account, autobiographical writings are *fragments* of lives. Fragments – as short as a phrase (*biographème*) or as long as, well, a multi-volume collection – are assembled for a purpose. They can trace connections within an apparatus and show how, for example, families are involved in a treatment decision about a infection in a hip prosthesis for an unconscious loved one.

For feminist geographers, even though exploring the personal lives of subjugated and marginalized people is commonplace, the presence of autobiographical details in a piece of academic writing still feels more intimate than reporting on someone else's private life. Research as an academic practice legitimates exposing people's lives in print, a practice supported through a formalized relationship as researcher and research participant, and cemented through a signed letter of consent. In autobiographical writing, there are no signed letters of consent between the researcher and the writer. Personal details are stitched into a text much like how information from interviews populate a research article. But the revelations seem raw and exposed, a little too personal. Readers are pressed into serving as an author's confidante, receiving private information that once known cannot be unknown. Part of the role of a confidante is not to tell, not to repeat particulars to anyone. And so those revelations are not repeated among academic audiences. Perhaps the reader may know the author, and may be embarrassed to mention such intimacies at their next meeting. Perhaps drawing out the juxtaposition of rational analytics and personal experience still jars academics, even feminist ones. The intimacy generated through this asynchronous engagement wraps what was written in a cloak of silence, much like keeping confidences in the realm of the unsaid.

When using one's own life as data, there is a tension between what details are available and what needs to be written in order for a reader to understand a specific point. There is no doubt, revelatory writing can be mind-numbingly tedious; details of one's life do not make for engaged reading. The writing seems stock, clichéd and glib, and often stripped bare, dismissive of the fulsome context of the recorded event. In reviewing my journal in preparation for this chapter, I filled my head with the names of nurses with whom I spoke, what I could find to eat on a restricted vegan diet with food allergies, and an endless array of blood gas readings. I reviewed what phone calls I made, when I made them, who phoned me, the times and duration of each call and what specifically transpired in each conversation. I was surprised at the words I had written in the margins depicting my mood or a specific feeling I had at the time. I relived each moment, many of which were unpleasant, deciding whether the moment facilitated a tracing or whether, if included in the analysis, would be extraneous information. I came to see the monotony of my notes, the pedestrian way in which I recorded how I manoeuvred through the complexities of care in the ICU. And I was reminded that the commonness of these acts demonstrates precisely what constitutes micro-connections in an apparatus that is so overwhelming on first encounter.

The vignettes

I use a hybrid auto-method of autobiographical writing with autoethnography. I write about a period of over a couple of weeks in the spring of 2010 through a set of four vignettes. I name people to facilitate reading my writing and your reading, but use pseudonyms to maintain anonymity. Equipped with a framework that permits me to draw out a handful of micro-connections within a complex institution and a method that can show both the complicatedness of writing autobiography analytically and the tireless triteness of disclosing intimate moments of everyday life, I tease out some of the intimacies of care around dying I found in an ICU. In contrast to much of the feminist geography literature on intimacy around the scale of the global intimate in geopolitics and political ecology (e.g. Mountz and Hyndman, 2006; Elmhirst 2011; Pratt and Rosner, 2013), the intimacies I speak of are those arising from family connections, emotional ties and bodily transitions. As the chosen designate to make decisions about my mother's healthcare, I became the one to liaise with her family physician, to both arrange for and provide hands-on care and to report to my brothers about decisions. Emotionally I was somewhat stretched personally and professionally, being distant from my brothers, estranged from my mother's family and dealing with workplace challenges. I had to determine the course of care, including the moment of death, for this frail woman who had been a key part of my life. Together these vignettes speak to what sometimes goes unsaid in the texts written for publication, trace micro-connections within institutions and disclose the specific slice of matter where I dwelt in those particular moments.

On the edge of ICU, 19 March 2010

Dr Plumber gave me a report over the phone: water in the lower lobe of one lung and oxygen seems to help. I asked about the BiPap machine – does the hospital have one or should I bring hers? Dr Plumber first said it does not matter for it would do no good, but then relented and suggested bringing it in from home. I also requested the addition of a gerontologist on the consulting team, a request in principle that has to be honoured but in practice it is more of a hit-and-miss affair.

I enlisted the help of my brothers to ensure that she wear the BiPap overnight. One brought it in, the other tried to get her to wear it. I phoned and spoke to Emily, the overnight nurse, noting its importance. Emily said that she couldn't actually do anything because the BiPap was not prescribed. I told her the story of how the stroke had affected her breathing and how she needed to get rid of the CO_2 overnight. She said, no, she wouldn't do it and then hung up.

The next day, I spoke with Dr Plumber. He said that she was getting more delirious. I asked again about the BiPap and the gerontologist consultation. He talked over me, dismissing me. There was no need for a BiPap; there was no record of stroke-impaired breathing. There was no need for a gerontologist; he knew what to do. Without a chance for me to respond, he abruptly hung up.

More phone calls. I phoned Dr Viscaine to see if she would review my mother's case, but she had no privileges at that hospital and she could not give them any information unless Dr Plumber requested it. I decided to contact the case manager, who was in charge of coordinating consultations and liaising with family members. So I phoned the nursing station for a name and number, and Bernadette refused to give me the number or tell me who the case manager was.

I tried accessing the case manager through patient advocacy, and Violet phoned back within five minutes. Violet talked to me off the record. She told me that I had been tagged as being difficult to reason with. Why? Because I asked too many questions. By the doctor? Yes, and so the nurses won't talk to you. I told her about the stroke and the BiPap, and asked that they request records from Dr Viscaine. She said that the doctor won't request the information because 'he doesn't like you'.

Five days later and only after an intervention by Directors in Aging and in Hospital Medicine within the hospital system, a gerontologist, Dr Belmonte, was assigned to my mother's case. She cleared out the communication lines: the nurses had to speak to me, my phone number was corrected (the wrong number had been recorded in the chart, which I still have difficulty believing), my brothers and I were the only ones permitted to access information about her medical condition, and she would personally contact Dr Viscaine. Fine. She then said that she understood that it was difficult to be far away and caring for a mother. I hesitated, not knowing how to respond to the shift in topic. Just short of being flustered, I responded by saying that it had never been a problem until this hospital visit, and that I was deeply dissatisfied in the way that this ICU functioned.

ICU permissions, 24–25 March 2010

A new physician rotated in, Dr Jaeger. Rather than improving, my mother steadily worsened. Blood gas readings showed CO_2 had crept into the 90s, when values usually were between 35 and 40 mm Hg ($PaCO_2$), and oxygen fell to just under 60, when values usually were between 75 and 100 mm Hg (PaO_2). More and more acidity. She was weak and couldn't sit up. Her breathing was shallow. I phoned the nursing station for updates. I meticulously recorded what the nurses read to me – Angie, Tasha, Dot, Kavlyn, Rena. My brothers were onsite, who gave me reports on her capacity to talk, her appetite, her general appearance and her ongoing hallucinations. I reported to Dr Belmonte, telling her that the hallucinations were like the ones she had when she has too much CO_2 in her lungs. If she were only on the BiPap, she would improve considerably, and then they could deal with the pain from the hip. She needed to be able to breathe on her own.

The next morning, my mother lost her breathing reflex and had to be intubated. I was not consulted; I was told only after the intubation. I was aghast. How could something as simple as needing to wear a BiPap get so out of hand?

I spoke with Dr Jaeger, again. I told her that the treatment has been appalling and that she is to get permission for all medical procedures and medications from me. She agreed that for future medical interventions, she would contact me to get permission. Then she went through the list of drugs. Okay, okay, what? Haldol? Dr Plumber had prescribed Haldol because her hallucinations indicated dementia. I thought you okayed it through the gerontologist. No, absolutely not. Morphine? No wonder she lost her breathing reflex! Water in the lungs, low O_2 and high CO_2, and you treat it with Haldol and morphine? I ended the call by saying that I take my responsibilities seriously and in order to make informed decisions, I need accurate information and that she was to provide that to me. She agreed. I sensed her agreement was offered begrudgingly.

Death and dying, 26–27 March 2010

I finally saw her. She was asleep. I sat beside the bed, letting her right hand rest on mine. For an hour or so. Everything was calm. The rhythmic ventilator made her breathing sound like a meditation tape. She began waking up. Opening her eyes, she caught a glimpse of me, and grabbed my hand with both of hers. She tried to talk to me. We quickly figured out a way to communicate: I would ask her questions, and then she would spell words into my hand. She tired quickly, but not before indicating that I should eat something.

I saw Dr Jaeger that afternoon. She dropped into the chair, crying. Between sobs, she told me I had to let her die. She told me that the ventilator was painful, and that my mother didn't deserve to be in pain. But I need more time. She told me I had no more time. I turned to my mom, and asked her if she wanted me to turn the ventilator off. No reply. I asked, do you want to keep it on? She squeezed my hand. I turned to the doctor and said the ventilator stays on. She told me that this wasn't enough: she was non-responsive and needed to die. I quietly asked her to leave. Late in the evening, the respiratory therapist, Eugene, whom I met before through a neighbour's kin, came to check the ventilator. He asked if I had spoken with Dr Jaeger. Yes. So, are you going to pull the plug? I paused, tilted my head to the side, and said, 'not yet' as firmly as I could muster.

The next morning, I wandered around the hospital looking for something to drink. I happened to take a wrong turn which brought me to the corridor full of offices. I knocked. A voice invited me in. He was the Director of Operations. He called the Director of Nursing over to join us. I told them my story. They said they would look into it. Brian retreated to his inner office while Evelyn accompanied me to the ICU that moment.

She skimmed the charts and said there clearly were problems. The overnight nurse had written that I had gotten her to respond. Evelyn concluded that there's 'someone in there'. She asked what I wanted to do. I said 'remove everyone associated with her hospitalization up to this point, and leave the good nurses'.

Inadvertent malpractice threats, 30 March 2010

Per protocol, or so I was told, when the family requests a change in attending physician, the ICU automatically sets up a family meeting. My brothers and I attended, along with our spouses/partners. Five healthcare professionals were there: the new attending physician, Dr Lebel; the charge nurse, who was the only one with papers in front of her; a respiratory therapist, who looked as if she were a recent graduate; Dr Belmonte, the gerontologist; and an ICU nurse, the woman who would actually follow my mother's progress for the upcoming week.

The room where we met appeared to be an old break room, filled with metal folding chairs systematically placed around generic, faux-wood seminar tables. I sat in the middle of one side, flanked on both sides by family. All my papers were stacked neatly in front of me, a checklist of items to address on top. The professionals sat on the other side, with their backs to the only door in the room. Except for the charge nurse, they sat at ease, leaning back with their legs under the table, crossed at the ankles. Their occasional laughter reminded me that this was their place of work.

The gerontologist opened with a couple of statements: 'This is a difficult time for all of you. We hope you are finding the emotional support you need.' I wrinkled my forehead, and glanced to either side of me. I was ready this time. 'Thank you, but we're able to find our own support systems. We don't see that our well-being is your responsibility.' I paused, looked down at my notes and began talking. I spoke of the commitment we had as a family to follow my mother's wishes with regard to living, being ill and dying. I said that these were laid out in her living will and durable power of attorney documents that have been on file in the hospital since treatment for her stroke years before.

I continued. I had sought advice about how to invoke existing reporting lines as power of attorney over healthcare decisions from a lawyer I saw that morning. At my mention of the word lawyer, they all quit slouching, placed both feet on the floor and leaned in over the table. The charge nurse sat up even straighter. It took a moment before it clicked: even though they knew they were there to clean up someone else's mess, they hadn't realized there was the potential for a malpractice suit.

Some reflections on setting up the invisibility of autobiographical parts

As tracings, these vignettes about an ICU begin to expose the bindings that keep embodied apparatuses glued together so that healthcare continues to function. Whether tightly compacted or loosely suspended, the elements interact, lying inert until called to action in particular situations. Once mobilized, things start shifting: information leaks through a conversation with a case manager, medical conditions disappear without internal records, demands in light of mistakes alter communication and bodily comportment shows what is important in the practice of care. The limited number of connections made in these vignettes show only how one person's healthcare came together in one ICU stay. Yet these connections, although singular, are not unique. They provide insight into how intimacies of caring and dying get taken up by family members and ICU personnel through particular protocols and practices in place for medical treatment. Though the vignettes invite further scrutiny, the texts themselves promote a particular type of analysis, one that already has been produced from a critical vantage point. And while the choice of what to include in the writing necessarily excludes other things, there is still the option to write about things that are usually excluded in analyses of healthcare in ICUs.

Tracing two requests, for the BiPap and for a gerontologist, draws out conflicting approaches to healthcare: the medicalized form (epitomized by Dr Plumber) and a holistic well-being form (represented across me, my brothers, the sitters and in part by Drs Belmonte and Lebel). The turf Dr Plumber was claiming was associated with anything having to do with the science of medicine. Biomedicine as the particular knowledge system organizing the ICU supported his thinking, his talking and his actions, even if it meant dismissing gerontology as a non-medical branch of medicine and physicians in different (and economically-competing) organizations of healthcare delivery systems (Loeppky, 2014). Dr Belmonte sought to have input into how elderly people may have varying reactions to conventional ICU medical practice, such as responding more slowly to treatment, having a heightened sensitivity to usual medication doses and being more attuned to death and dying. Dr Lebel pursued an objective set through consultations with me as my mother's legal representative and the gerontologist: to go home to die.

In decision-making around death and dying in an ICU, I was not privy to most of what was said or what went unsaid among those working in ICU. I had no ongoing access to my mother's chart and only engaged the physicians and ICU staff in fits and starts. A key insight was through Evelyn, the Director of Nursing, who reported to me that the physicians were on different tracks. The ones in ICU were treating my mother to die and the gerontologist was taking a wait-and-see approach. Once the personnel changed, the nurses and physicians talked with me and showed me her chart. I could see the record of their practices. I was also on-site and saw the ICU staff in action, taking vitals, giving medications and swabbing my mother's mouth.

Life outside the ICU for the staff came through only in bursts: Dr Plumber's adamant refusal to be influenced by a physician he considered not to have appropriate medical training, Dr Jaeger's breakdown over the thought of pain from a ventilator tube and the respiratory therapist's presumption of feeling close enough to me to ask about pulling the plug. I knew less about Drs Belmonte and Lebel, suffice to say that Dr Belmonte seemed preoccupied with my emotional well-being and that Dr Lebel realized when we met that I was, in his words, 'calling the shots'. I know bits of what went unsaid from my side. I was preparing a talk about *Magnum, PI* with regard to nationhood, masculinity, psychiatry and veterans. I was attending phone meetings with my colleagues to prevent the closing of the programme I taught in. I was talking daily with various members of a branch of a national bank in another state to amalgamate my mother's accounts. I was tussling with my own ongoing chronic illness teetering on the edge of another collapse around fatigue and pain. I was re-learning how to communicate with my brothers as adults and not as the adolescents I remembered them to be when we had last spoken in earnest.

These vignettes arise out of a hybrid approach between autobiographical writing and autoethnography. Writing them here as tracings resituates assembled acts into the said category, but only for a handful of people. This hybrid approach permits me to open a discussion of auto-methods being useful not only in tracing some micro-connections within an apparatus of healthcare expressed through an ICU of a teaching hospital in a southern US city, but also in mapping out how decisions about death and dying are not extraordinary or discrete events. Writing about my situation in this particular locale in relation to care in the ICU allows me to tease out intimacies of care in order to understand competing readings of healthcare practices with regard to, for example, lay knowledges, biomedicine, medical and non-medical physician care, legal and ethical responsibilities, and the emotional attachments of love. Intimacies of care in the ICU are not solely composed of doctors' orders and nursing records; rather, they are a delicately balanced set of fragile combinations of biomedical decisions, family experiences, emotional attachments, affective atmospheres (Anderson, 2009), communication methods and styles of interaction that signal some of the complexities about bodies, illness and dying. These complexities can effectively be accessed through auto-methods, ones that can show that although this is my story, it is not just my story. It really is more of an 'it-them-me-you-here-me-that-you-there-her-us-then-so' (Cook, 2001, p.120) tale of tracings and micro-connections in a 'specific slice of matter' (Braidotti, 2013, p.35).

Epilogue

Drs Belmonte and Lebel continued to work with me. We went over in detail my mother's complex history of stroke and infection; we discussed how differences in lay and medical definitions of life support clash when it comes to making decisions about invasive and non-invasive measures; and we reviewed what medical plans were on offer to permit my mother to go home. Dr Belmonte reminded me that the first pressure trial had lasted only three minutes, and my mother couldn't even assist in breathing. Dr Lebel noted that the lungs are still impaired but thought things might be different if she were not taking medications that impaired her breathing. So, on 1 April, the second pressure trial, planned for ten minutes, began. Oxygen saturation 100 percent; blood pressure low, but within range; heart beat fast, but not excessive; pulse steady without interruptions. The next day I asked when the third trial would take place. Dr Lebel said that there was to be no third trial for she was doing so well that they would keep her off the ventilator. She had been breathing on her own. If, after another 24 hours, she was still breathing well enough, they would extubate. She was and they did.

Six years later, my mother still wears the BiPap – at night.

Works cited

Anderson, Ben, 2009. Affective atmospheres, *Emotion, Space and Society*, 2(2), pp.77–81. doi:0.1016/j.emospa.2009.08.005

Arrigio, Bruce, 2002. *Punishing the mentally ill: a critical analysis of law and psychiatry*. Albany: State University of New York.

Besio, Kathryn, 2005. Telling stories to hear autoethnography: researching women's lives in Northern Pakistan. *Gender Place and Culture*, 12(3), pp.317–32. doi:10.1080/09663690500202566

Bondi, Liz, 2014. Feeling insecure: a personal account in a psychoanalytic voice. *Social and Cultural Geography*, 15, pp.332–50. doi:10.1080/14649365.2013.864783

Braidotti, Rosi, 2013. *The posthuman*. London: Polity.

Butz, David and Besio, Kathryn, 2009. Autoethnography. *Geography Compass*, 3(5), pp.1660–74. doi:10.1111/j.1749-8198.2009.00279.x

Cook, Ian, 2001. 'You want to be careful you don't end up like Ian. He's all over the place.': autobiography in/of an expanded field. In: Pamela Moss, ed., *Placing Autobiography in Geography*. Syracuse: University of Syracuse Press. pp.99–120.

Elmhirst, Rebecca, 2011. Introducing new feminist political ecologies. *Geoforum*, 42(4), pp.129–32. doi:10.1016/j.geoforum.2011.01.006

Foucault, Michel, 1980a. The confession of the flesh. In: C. Gordon, ed. *Power/Knowledge: Selected Interviews and Other Writings, 1972–1977*. New York: Pantheon Books. pp.194–228.

Foucault, Michel, 1980b. Truth and power. In: C. Gordon, ed. *Power/Knowledge: Selected Interviews and Other Writings, 1972–1977*. New York: Pantheon Books. pp.109–33.

Kobayashi, Audrey, 2009. Situated knowledge, reflexivity. In: Rob Kitchin and Nigel Thrift, eds., *International encyclopedia of human geography*. Oxford: Elsevier. pp.138–43.

Loeppky, Rodney, 2014. *Accumulation and constraint: biomedical development and industrial health*. Halifax and Winnipeg: Fernwood Press.

Longhurst, Robyn, 2012. Becoming smaller: autobiographical spaces of weight loss. *Antipode*, 44(3), pp.871–88. doi:10.1111/j.1467-8330.2011.00895.x

Moss, Pamela, ed., 2001a. *Placing autobiography in geography*. Syracuse: Syracuse University Press.

Moss, Pamela, 2001b. Writing one's life. In: Pamela Moss, ed., *Placing autobiography in geography*. Syracuse: University of Syracuse Press. pp.1–21.

Moss, Pamela, 2013. Becoming-undisciplined through my foray into disability studies, *Disability Studies Quarterly*, [e-journal] 33(2), http://dsq-sds.org/article/view/3712/3232 [Accessed 16 November 2016].

Moss, Pamela, 2014. Some rhizomatic recollections of a feminist geographer: working toward an affirmative politics. *Gender, Place and Culture*, 21(7), pp.803–12. doi:10.1080/0966369X.2014.939159

Moss, Pamela and Prince, Michael, 2014. *Weary warriors: power, knowledge and the invisible wounds of soldiers*. New York: Berghahn.

Mountz, Alison and Hyndman, Jennifer, 2006. Feminist approaches to the global intimate. *Women's Studies Quarterly*, 34(1/2), pp.446–63.

Pratt, Geraldine and Rosner, Victoria, eds., 2013. *The global and the intimate: feminism in our time*. New York: Columbia University Press.

Richardson, Laurel, 2011. Hospice 101. *Qualitative Inquiry*, 17(10), pp.158–65.

Sharkey, Judy, 2004. Lives stories don't tell: exploring the untold in autobiographies. *Curriculum Inquiry*, 34, pp.495–512. doi:10.1111/j.1467-873X.2004.00307.x

Smith, Dorothy, 2005. *Institutional ethnography: A sociology for people*. Lanham, MA: Altamira.

Valentine, Gill, 1998. 'Sticks and stones may break my bones': a personal geography of harassment. *Antipode*, 30(4), pp.305–32. doi:10.1111/1467-8330.00082

19 Places of the open season

Sarah de Leeuw

Prologue

This is an ethnographically-informed creative non-fiction essay that dialogues and works with trends in critical anti-racist feminist geography addressing colonial violence and Indigenous peoples and spaces. The essay contemplates remote and overlooked places in Northern British Columbia (Canada) located along Highway 16. Highway 16 is colloquially known as 'The Highway of Tears' because of the 18 (or more) mostly Indigenous women who have been murdered or gone missing along its shoulders. Drawing on the literary flexibility offered by creative non-fiction, and on more than two decades of working and publishing in the genre, I use multiple textual registers in this essay, from poetic language to research interviews, from staccato sentence structures to sound assemblages, from other academic texts to excerpts from popular media. This assemblage style is meant to evoke the variety of emotional and material places making up the always colonially-impacted geographies along Highway 16. These places include homes of on- and off-reserve Indigenous families whose children are routinely removed by the provincial government, women's centres and police detachments staff tasked with organizing search parties for women's bodies dumped in ditches, and vast regions of forest and watersheds routinely characterized by both industry executives and often urban dwelling nature-enthusiasts as unpopulated and open for development and discovery. Ultimately, these geographies are also the ones I call home, having grown up and spent the vast majority of my life in Northern BC and along Highway 16.

Part I

Here is what my friend Mary, Gitxsan-Dakelh Hereditary Chief and Matriarch, says to me after I have pulled myself up (way up) and into the cab of her shiny black Special Edition Ford F-350.

What she says is said as we are driving home from the airport, down the highway over the huge steel and concrete bridge that wings itself across the confluence of the Nechako and Fraser Rivers, two of British Columbia's largest rivers, rivers that meet on the outskirts of Prince George, a city where both Mary and I live.

Remember that Dakelh translates into 'people who go around by boat'. Remember that the truck is like a huge boat, that we are crossing rivers. Travelling.

She says to me: 'When me and everyone else were standing there in court a few months back, when they were reading those details about how he killed the women, how he wore the same set of clothes to kill all four of them, how when totally accidently and because a police officer had nothing more than a funny feeling about a truck pulling out from a small side road, and for nothing more than being caught for speeding, he was pulled over in the middle of a winter night onto the side of the highway and there was blood everywhere and on him, and, well, he said he'd just killed a deer, beating it to death with a pipe for fun. And the police officer almost believed him, you know? Except it wasn't hunting season. So the officer had a second thought.' Right then her mother turned to me and asked, 'Mary? Mary? Is that how my baby girl died?'

Voice.

Gives out.

Beyond tears.

A daughter will always be a mother's baby.

Windows open.

Mary says: 'We live in an open hunting season. This is not a place. We are no place. We are a season, an open season.'

The top end of the bridge we have just driven over?

One of three places where bodies of northern women have been found. Found dumped. Found slaughtered. Found animal-chewed. But. Found. Two First Nations women.

One legally blind fifteen year-old girl.

In total, four. Four. Four once alive breathing cell water breath finger digit bridge of nose life daughter. Four.

Quickly. Less than two years. Like a frenzy.

One body, one woman, never found.

Three of the women, mothers. One, a mother of six.

The ground should shudder. The ground should shudder. The ground shudders.

Let us go by boat.

Let us go.

Part II

Summer run-off. Water is high. Cotton wood trees growing close to the banks are sliding into the river, shore erosion.

The truck is hot.

Once, and for thousands of years, giant sturgeons swam in the waters Mary and I have just driven over. Prehistoric scales in current.

Then, in 1951, construction on The Kenney Dam began. Almost 100 meters high, more than 450 meters long, 3 million cubic meters of material: until 1960, the largest rock-fill dam in the world. Power generation for an aluminum smelter.

To cool the newly damned waters, two spillways are constructed. What was flooded, never recovered.

Almost a decade ago, Mary introduced me to a ranchin' man: cowboy hat, boots, mustache, Wrangler jeans. Hands made for holding reins. He lives on the edge of those spillways, ranging cows in the wreckage, telling stories. More than 70 years after that dam got built, coffins and burial boxes from the Cheslatta T'en people still wash ashore on the beaches of reservoir lakes in the western reaches of the Nechako River watershed.

The Cheslatta T'en people, also of the Dakelh Nation, tell their story like this: in 1950, when the world of Canada was post-war bungalow and refrigerator booming, when in remote Northern British Columbia it is the height of berry picking and salmon fishing season, Indian Agents arrive in their village in a valley the people have been living in for more than 10,000 years. They have always gone by boat. The water that will come cannot be escaped by boat.

The Indian Agents arrive with papers and cheques, x's already penned in as agreement for the $77/hectare the Indians will get, compensation for moving on. Down river, non-Indians are getting $1,544/hectare, with moving expenses paid.

The Indian Agents tell anyone who's listening, mostly children and the oldest of people who cannot travel and who are not on summer trap-lines or picking berries or fishing for salmon, that the village will be gone in 10 days.

A flood will come, say the Indian Agents; a flood will take your children if you don't move on.

Cheslatta T'en underwater.

A flooding season.

The people who go around by boat are offered no boat, are given no way around. The effort is made for them to be gone.

Here is another part of the story.

A young Cheslatta T'en man returns from his summer trap-line and there is nothing but water where once stood every single thing he loved. Garden. House. Graves of ancestors. A tin from his grandfather. A small piece of wood he'd carved as a boy.

The waters of the Nechako River run south and east from Cheslatta T'en to Prince George.

In 2010, just up hill from the riverbank where now warmed and sturgeon-free Nechako waters empty into The Fraser River, the remains of Cynthia Frances Maas are found.

As everyone driving into Prince George from the airport does, Mary and I drive by the place Cynthia Frances Maas was found. Mary turns and says to me, 'When me and everyone else were standing there in court a few months back . . .'.

I think: a flood will take your children. A hunting season. Damming. Indians. Moving. Soft shoulders of highways, side roads and gravel roads and trap lines and rivers in Northern British Columbia.

Think of Kathleen Stewart's *Ordinary affects*, how she writes that 'the potential stored in ordinary things is a network of transferences . . . it can be as palpable as a physical trace' (2007, p.21). How she says that 'ordinary scenes can tempt the passerby with the promise of a story let out of the bag . . . matter can shimmer with . . . undetermined potential . . .' (2007, p.23).

Think about the ground, about water, about people and about potential. About a boat. About crossing a river. About something that might one day shimmer.

Part III

The flight from Vancouver to Prince George, British Columbia's 'Northern Capital', takes just over an hour. I always think the paint on the plane shimmers. I do. I think of it shimmering, like a sky-boat lifting up up up above the horizon, its fuselage belly the stomach of a shimmering fish, invisible to predators in deeper water looking up.

In some ways, leaving for Prince George from Vancouver is not unlike leaving one country for another. Vancouver's sleek high-rise condos, quaint pottery shops, yoga studios, early-April cherry tree blossomings, steady streams of commuter-cyclists, and a municipal Mayor who made his fortune with *Happy Planet* organic fruit and veggie juices.

Northern British Columbia, a region larger than France, has lower life expectancies than the rest of the province, larger numbers of people living with type 2 diabetes and obesity, greater rates of tobacco and alcohol use, more significant rates of domestic violence and, for three years running and because of gang shootings and street stabbings and murders of survival sex-trade workers in Prince George, the city was named Canada's most dangerous city by the country's national magazine *Maclean's*.

These metrics say nothing of the less quantifiable differences. Things about all the people who own huge matted-fur mutt-husky dogs or scarred unneutered pitbulls, who know how to skin moose, tan hide or smoke salmon: the amount of mud and rust on vehicles that are so often diesel-engine trucks with split and cracked-up windshields, quads or chainsaws in the back; the giant billboards on primary highways reminding women how many of us have been murdered or gone missing while hitchhiking; the spruce trees or red cedar trees with sometimes more than fifty bald eagles dotting their bows, each predator bird waiting for that precise split-second to rocket down into schools of twisting silver oolican fish running fast and strong in some of the province's mightiest rivers, some of the largest watersheds in the country.

To an outsider, we might seem to know each other well, those of us flying north on the small jets, those of us who make the flight often, who have lived for decades or grown up in Prince George or in one of myriad smaller communities scattered across the north-central region of the province to which Prince George serves as the place with the most central airport.

Those of us who seem to know each other well are not the gently rounded baby-skin-smooth-close-shaven German, Swiss or Texan men in their late-50s wearing the latest in architecturally-crisp eye-glasses and hand-tooled leather designer-made belts that they will tell you their wives gave them for Christmas last year. They will tell you this after buying no fewer than four miniature bottles of scotch and two bottles of premium import airline wine.

These are the tourist men, the men who are spending 10 days with the guys, men in camouflage-patterned puffy down jackets or Arc'teryx Gortex and moss-green base-layers, in polarized sunglasses. Men with high-powered rifles or lightweight fly-fishing rods, men who are flying north with grizzly-bear kill-tags in their pockets, heli-hunting lift-passes to shoot mountain goats, creased maps with highlighted roads showing exactly where to drive rented 4x4 trucks to fishing lodges in remote inlets on the edge of wildernesses named things like The Great Bear Rainforest.

Those of us who seem to know each other well are also not the younger men, transient quiet, almost-boys who make up a boom workforce, under-35-year-old physiques still taut and toned but with eyes blurry from a week of binge-drinking and hard-drugging in southern or eastern Canadian cities, beer and rye whiskey and maybe some meth or coke sweating out from pores in their slumped-down for airplane travel bodies.

These young men are wearing pilled, stretched XL black polyester sweatshirts with 'No Fear' or 'Tap Out' or 'MMA Fight Real' scrawled on the chest. Their flat-billed truckers' caps or their toques are pulled down low, just like their thick-canvas Carhart carpenter jeans bunched over their unlaced steel-toed work boots, their dirt-encrusted workplace safety-stickered orange hardhats resting like giant jock-cups on their laps.

They are the boys flying north to work in mines, in logging camps, on pipelines and paving crews and fracking drills and explosive charges, in trucks and on trains that carry coal, on the veins that open to spew natural gas ready for liquefaction, ready for shipment to places far, far away.

They are the young men of Northern British Columbia's boom, young men clearing the equivalent of $120,000 annually who are sometimes paying more than a thousand dollars a month for a shared bedroom in a house with seven or eight other workers not so different from themselves in towns like Kitimat, built from scratch to house Alcan workers in 1951, an entire community dreamed into existence by Clarence Stein, visionary of the Garden City Movement in the United States who ensured cul-de-sacs were necklaced with small island oases of greenery and flowers, who sketched out for Northern BC a gently curving net-work of neigbourhoods and walking paths for imaginary company wives, mothers with babes meandering to and fro from two-storey suburban bungalows to school-yards and a small central shopping district as their men worked for the aluminum smelter, powered by the Kenney Dam.

Remember the man who returns from his summer trap-lines. Remember the cemeteries of ancestors.

Think of losing everything in 10 days.

I am telling you a story. There is a boat. There are women. There is a shim-mering.

Remember.

Think of the tin, small so that it might be warmed in the palm of a hand when held, given to the man by his grandfather. Think of hands shaped into cradles, curved for the slip of leather reins, a ranchin' man, hands slowly calcifying with rheumatoid arthritis, something that cripples so many northern First Nations for reasons unknown.

Those of us who, to an outsider, might seem to know each other well offer that impression because there's a sense of knowing one-another's stories. I am telling you a story.

We introduce ourselves by where we're from in the north, by what small place we went to school, what precise location we first fell in love and, within moments, we are finding connections to people we have in common. We Northern British Columbians tell each other stories, stories about our children, our wives and husbands and cousins and grandkids and uncles. We tell stories about bits of local news, like the scandal of some small town mayor, how many black bears were shot in people's backyards, what was stolen from porches during ravens' nesting season when all the big black birds go crazy for anything shiny, or sometimes about places where berry picking is good that year or about progress on getting health care professionals up into our no-road towns or a local court case we are all following but which is never featured on provincial or national news, our stories invisible to the rest of the country, our places strange no-places except to the small numbers of us who live in Northern British Columbia and call it home.

Our stories travel, but never too far it seems. And always, always, they come back to us. We chatter with each other as if we all know, or are at least familiar with, each other's stories: our stories orient us to each other, to the vast regions of no places, what Mary calls a season, an open season.

Like Rebecca Solnit writes, stories are all in their telling: 'stories are compasses and architectures: we navigate by them . . . which means that stories are geography . . . a way of traveling from here to there' (2013, p.8).

Or, as Cherokee-Scottish author Thomas King would say, 'the truth about stories is that that's all we are' (2003, p.2).

I am telling you the stories that touched a Cheslatta T'en ranchin' man, that reached out to Mary, that touch down on me. From here to there. From me to you. The promise of a story let out of the bag . . . matter can shimmer with . . . undetermined potential . . . .

Mary and I are each other's stories, the stories we share. It is true that we know each other well, that we have told stories to each for more than a decade. If stories are geography, if stories are all we are, a sort of ethno-biography, a thick accounting, an exploration, a narration, all bundled tightly together and then told and received – as Rebecca Solnit would have it, 'a hybrid style . . . with permission to wander . . . journalistic, critical, lyrical . . . first person experience . . . investigation, analysis and description' (2013, p.12).

Mary and I are partly each other's geographies, each other's ethnographies, architectures, bodies and histories – storytellers of each other and ourselves, a sort of mapped auto-geo-ethnography. We are partly each other, but we are also ourselves still untold, a story in the making.

Our stories touch in such great part because we are women working with women living in Northern British Columbia, women driving the highways, today together, but so many other times apart, the horizon ahead where hay field, pine-beetle killed forests, lake, Indian Reserve, fishing weir, mine, clear-cut, snow bank, logging truck, hitchhiking girl meets sky turns into story, a story we tell each other.

I am telling you a story.

Part IV

In a hotel ballroom, one of those anemic beige windowless spaces where people do everything from getting married and graduating from high school to hosting consultations about waning sockeye salmon stocks, Mary and I are sitting in a circle of hard chairs, arching brass backs holding in place now stained fabric, the kind of chair that once tried to pass for elegance. We are in a circle of women from across Northern British Columbia. On our laps is a report entitled *Forsaken* issued in 2012 by the Honorable Wally Oppal. The names of murdered and missing women, many from Canada's poorest postal-code (DTES, i.e. Vancouver's Downtown Eastside) but many too from Northern British Columbia, and so so so many of them Aboriginal women, names printed in red against the black text: the names pop out as we flip through the pages, turning to the 63 recommendations.

Rebbeca Guno, Nisga'a First Nation from Northern British Columbia.

Olivia William, born in Burns Lake from the Lake Babine First Nation in Northern British Columbia.

Alberta Williams. Found 37 km east of Prince Rupert, BC, near the Tyee Overpass, strangled and sexually assaulted.

Aielah Saric Auger. Age 14. Found dead in a ditch on Highway 16 near Tabor Mountain, 20 km east of Prince George.

In the two years since the report was released, not a single recommendation has been implemented.

Sitting between Mary and me, the sister of Ramona Williams.

Ramona Williams, hitchhiking to her friends' home in Smithers, Northern BC on 11 June 1994. Age 16. Remains were found just under a year later near the Smithers Airport. No charges ever filed.

Her sister is wearing long, beaded earrings, woven through with pheasant feathers: black jeans, nails bitten to the quick. She says: 'I miss my sister every day.'

Mary and I look at each other, hold each other's gaze.

Body, sky, sky, body. Sister. Skin. Smithers sky, skin story.

Before Mary's and my stories touched, I am working in a women's centre in Terrace, a tiny community in Northern British Columbia on the edge of the Skeena River, when a woman walks in. She does not say hello but, instead, says: 'My sister's been missing for two years and no one's done nothing. What are you people doing about it?'

Lana Derrick. Age 19. Last seen at the gas station where I filled my tank that very morning, grey slush roadside snow, early winter. The billboard-sized photos of her on the highway's soft shoulder show a serious young woman with large black plastic framed glasses resting on teenage-pudgy cheeks, a black ribbon choker necklace with a heart locket resting on the tender spot where trachea meets larynx. Around her photo are emblazed neon-blue letters that read 'missing' and offer a toll-free hotline to phone in tips.

Lana Derrick is no place. She has never been found. A vanishing in a place that is not a place but is a season, open season.

By 'you people' Lana Derrick's sister means all the people whose jobs it is to protect women and girls: police officers, detectives, social workers, band-councillors, community outreach employees, drug and alcohol and addiction therapists, doctors, well-intentioned feminist front-line anti-domestic violence workers – none of us who are stemming the flooding, none of us who are putting a stop to hunting season.

Here is another story. Told by a woman named Mary Ellen Turpel-Lafond. She tells stories with the voice of a lawyer. She adds her voice to the voices of many others. Reports. Inquiries. Appearances. Press releases. One part of the story is Paige's. It's a story entitled *Abuse, indifference and a young life discarded.* Can you see the faint traces of wings across the report bearing baby pictures of Paige? Ovoid lines like eyes, eagle raven. Another part of the story is entitled *Lost in the shadows: how a lack of help meant a loss of hope for one First Nations girl.* And then, *Out of sight: how one Aboriginal child's best interests were lost between two provinces* and *Fragile lives, fragmented systems: strengthening supports for vulnerable children.* I wonder about all the stories, all the reports from 'you people'. From people like you and like me.

Part V

In the hotel room, so many years later, in the circle of chairs, in the circle of women with a report on missing women on our laps, Ramona William's sister begins to cry. She cries very, very quietly. Her shoulders move up and down, as if she is shrugging, shrugging, shrugging something away. She too is telling the world a story.

As we sit, as Ramona William's sister sobs, two other women begin speaking. They pick up the thread of a story that has already begun, a story stitched into an intersection of streets a few blocks south of the hotel, a story that is an attempt to make sense of things that make no sense.

Remember that those of us you are reading about live in a season, an open season. Our place is a place of open season.

'Can you imagine', says one, the other saying at almost the same time, 'imagine, imagine . . . So young.'

They turn their heads to the left.

Other women know where they are looking.

Towards the courthouse, the police station with the temporary jail cells. Where the trial of now 24 year-old Cody Alan Ledgebokoff, stopped for speeding three years ago and answering only that he had clubbed a deer to death, has just begun.

Twenty-one years old when he was arrested. Charged with killing four women. Two First Nations women. One 15 years old and legally blind, met online using his moniker *1CountryBoy*, something he once bragged that northern girls gravitated towards. The reason all the blood at first made some sense to the police officer who pulled Ledgebokoff over for speeding was the explanation: a deer clubbed to death for fun, augmented with the detailed observation 'it's what we rednecks do for fun in the north, right?' This latter detail was not out of place, at first, to the police officer who pulled Ledgebokoff over. It rang true, made sense.

In some ways, leaving for Northern British Columbia from Southern British Columbia, where 70 per cent of the province lives clustered with 250 km of Vancouver, is not unlike leaving one country for another.

As might only happen in this other country, in this huge region with a tiny number of people, when I host a dinner at my house in the middle of winter, two professional cowboy poets show up. These women can recite verbatim from memory poems that go on for 20 minutes at a time, all rhyming weaving backtracking flowing, all charting the trials and tribulations of men and wives, of love and losses, on trails and mountain ranges, with pack horses and whiskey in tow.

They tell stories as maps.

Mary and another close friend of hers from an isolated Indian Reserve also show up to my house that night. More stories begin. Stories get us from here to there. It turns out Mary grew up just down the road from the cowboy-poets, who lived beside the Ledgebokoffs: Mary's family on the Nakazd'li Indian Reserve, Cody and the cowboy-poets' families in non-Indian Reserve Fort St. James. Less than 1 km away from each other. The cowboy-poets speak about a sweet round-faced boy tossing a football in the family back yard. Mary works with the families of the daughters that round-faced boy is accused of raping and then beating to death.

Here is another northern story, which touches down on the same courthouse where, every day, Cody Ledgebokoff sits in a small plexiglass box being watched by the small numbers (under a dozen, 4 weeks into the trial) of people who bear witness to the trial. Family members of the missing and murdered women.

In early 2002, in the same courthouse where Cody Ledgebokoff sits, a small group of young Indigenous girls give evidence against local Prince George judge David William Ramsey. Ramsey is later convicted of and jailed for multiple violent sex crimes between 1992 and 2001 against at least five First Nations girls, yes, all First Nations girls and some as young as 12, who worked on the streets of Prince George. Most of the girls were wards of the state whose apprehensions from their families Judge Ramsey presided over in the Family Court room. Years later, when I speak to one of the girls in a park just downstream from where the body of Cynthia Frances Maas was found, she says to me: 'I [have] always felt lost. Like I [have] no home ... I never felt like I belonged anywhere. I always felt out of place.' Then she tells me how, as part of the compensation for the abuse she suffered at the hands of a judge, she is taken on a tour to visit the site where some evidence suggests her ancestral family was buried. She speaks of feeling something beyond words, a grounding. To see, even just possibly, fleetingly, where members of her family might have generations ago lived offers her a feeling she has never experienced. A feeling that she does have some place, some anchor, in this world.

These are the stories of who we are. We are our stories. I am telling you a story.

Sometimes Mary pulls her big black Ford 350 into my driveway and we go for a drive, no place in particular. Mostly we do not speak of the places we pass, places where the bodies of women have been found. Mostly we do not speak of the posters, taped to hydro-poles, with the faces of missing women and girls staring out rain-stained peeling slipping back to pavement and earth. We tell each other different stories, and we get from here to there, and our stories drift and settle and drift and move.

Once Mary introduced me to a ranchin' man. To stories of watersheds that stretch across Northern BC, aluminum smelters to graves to body-shaped indents on moss and pine needles, shoulders of highways, bridges over sturgeon-less waters. Once Mary stood in a courtroom.

Once I listened to a girl who told me her story of being assaulted by a judge who presided over that same courtroom.

In the truck with my friend Mary we drive towards the horizon, a road that meets sky. A season. A story. Here are names for the story. The stories of ordinary places, places many have never heard of.

Try to pronounce these names: Kitwanga, Wet'suwet'en, Gitsegucla, Skidgate, Metlakata, Chimdemash, Kleanza, Kitselas, Kispiox, Lheidli T'enneh, Hagwilget. Keep them on your tongue. Find them on a map. Find more. Say them out loud.

Tell this story.

Know we live in places of the open season.

We live.

Works cited

British Columbia Representative for Children and Youth, 2011. *Fragile lives, fragmented systems: strengthening supports for vulnerable infants. Aggregate review of 21 infant deaths*. Victoria, BC: Government of British Columbia. [pdf] Available at https://www. rcybc.ca/reports-and-publications/reports/cid-reviews-and-investigations/fragile-lives-fragmented-systems [Accessed 14 July 2016].

British Columbia Representative for Children and Youth, 2013. *Out of sight: how one Aboriginal child's best interest were lost between two provinces*. Victoria, BC: Government of British Columbia. [pdf] Available at http://cwrp.ca/publications/2757 [Accessed 14 July 2016].

British Columbia Representative for Children and Youth, 2014. *Lost in the shadows: how a lack of help meant a loss of hope for one First Nations girl*. Victoria, BC: Government of British Columbia. [pdf] Available at https://www.rcybc.ca/reports-and-publications/reports/cid-reviews-and-investigations/lost-shadows-how-lack-help-meant [Accessed 14 July 2016].

British Columbia Representative for Children and Youth, 2015. *Paige's story: abuse, indifference and a young life discarded.* Victoria, BC: Government of British Columbia. [pdf] Available at https://www.rcybc.ca/paige [Accessed 14 July 2016].

King, Thomas, 2003. *The truth about stories: a native narrative.* Toronto, Canada: House of Anansi Press.

Oppal, Wally T, 2012. *Forsaken: report of the Missing Women Commission of Inquiry.* Victoria, BC: Government of British Columbia. [pdf] Available at http://www2.gov.bc.ca/gov/content/justice/about-bcs-justice-system/recent-inquiries [Accessed 14, July 2016].

Solnit, Rebecca, 2013. *The faraway nearby.* New York: Viking.

Stewart, Kathleen, 2007. *Ordinary affects.* Durham, NC: Duke University Press.

Concluding remarks

20 Intimate research acts

Pamela Moss and Courtney Donovan

As the contributors in this collection have shown, writing intimacy into feminist geography is a multifaceted project. Included as part of the project are feminist geography researchers who write themselves into their research as part of their academic practice. The personal writing takes many forms, all with the intention of furthering a set of analytical concerns about power and the production of knowledge in the everyday. Tracing the effects of including their own stories as part of the research process is also part of writing intimacy into feminist geography. These contributors go beyond reflection as mere musings; they engage critically with the effects of entering the field, living their lives alongside their research participants and sharpening their acumen in analyzing what is going on in everyday life. Intimacy as a research topic needs attention in order to disentangle the ways in which embodied, emotional relationships get cultivated either on purpose or by happenstance as affinities or antagonisms intensify. Following intimate bonds with humans, nonhuman entities and non-living things serves as a means through which to explore issues that matter to feminist geographers.

Yet writing intimacy into feminist geography is not solely nor simply about the way in which a feminist geographer as researcher, academic or analyst generates intimate texts, reflects on the effects of including intimacy as part of research or utilizes intimate connections to examine structures of power. It is about uncoupling personal connections to see how they work, what they do and what they produce. Even though some of these connections as well as the uncouplings are cloistered, they are still crucial in the social organization of the everyday. Contributors show how intimacy is as commonplace and personal as it is rare and collective.

Across these pages are instances of intimacy that serve as discussion points on how to talk about intimacy methodologically. Present in each chapter is a loosening of the bindings that tie intimacy to a particular conceptualization. Intimacy has long been linked to holding a particular type of knowledge – the personal, the subjective – while autobiography has been the disclosure of that knowledge. But the contributors both individually and collectively show that this does not necessarily have to be the case. Although there are many autobiographical accounts, there are no solipsistic narratives that neglect the politically urgent need to dismantle systemic oppressive power structures. Contributors enter into commanding knowledge systems located in the state, institutions and cultural systems and challenge the authority they wield. What effects their contestations have may not be ascertainable now, in a few months or even in ten years. For now, we can say that they challenge power by examining the banalities of everyday life, such as caring for a loved one, going to the doctor and visiting a friend. The contribution is not an instance of the more humanist idea of change underlying, for example, the random acts of kindness movement. These chapters both individually and as a whole are part of a collective project of transformation (Braidotti, 2013), joining other acts carried out by numerous feminists that contribute to a much wider movement for change. Although the contributors write only about one particular aspect of their academic and political practices and not their entirety, their contributions provide insight into how intimacy works and what it can do in their support for more-encompassing strategies for change.

As a collective the volume opens up the form writing about intimacy takes, widening from the solid base of personal writing and autobiographical analysis to include creative bricolage, graphic memoir and creative non-fiction essay. We have come to understand these contributions to this collection as a set of intimate research acts, ones that *enact* writing intimacy into feminist geography. By enact we mean that their presence as written pieces is actually doing what we are claiming; that is, the contributors are writing intimacy into feminist geography. Our use of enact is similar to Mol's (2002, p.41) when she lays out her reason for preferring enact over performance, that is, not to be burdened by the conceptual baggage associated with performance. We also, like Mol, wish to keep the meaning of enactment open so that its flexible, porous and expandable attributes are refracted through further contributions in other venues.

As part of writing intimacy into feminist geography, we introduced the idea of muddling intimacy methodologically in the introductory chapter to describe what these contributions as a collection do. Muddling intimacy in research means accessing and attending to those things within the everyday, such as the intensities of emotions, affective cultural practices, bodily sensations and subtle shifts in meaning and routinized practices. And it is within resonances between entities and connections among people, nonhuman entities and non-living things that intimacy can be taken up and examined. How feminist researchers convey these entanglements through their writing is part of what liberating methods can do for knowledge production and social change (after DeVault, 1999; see Moss and Falconer Al-Hindi, 2008). In their writings, these feminists follow connections, links, boundaries, relationships, associations and correlations within everyday contexts they encounter as part of their research in the sense that intimacy provides a means through which to liberate methods and create intimate research acts to engage in. They constructively muddle intimacy methodologically by layering their accounts with insights about intimacy, about intimate contact with people, nonhuman entities and non-living things, and about how to write up their thoughts for sharing. Indeed, these chapters provide strategies – conceptual, theoretical, analytical – for sorting out how intimacy works, what it does and what it produces.

In the rest of this chapter, we do two things. We first frame our understanding of muddlings as intimate research acts within an affirmative politics. We then present reflections on some of the resonances we have found among the contributions. We note that as the contributors write intimacy into feminist geography, they take up specific academic practices that are themselves intimate research acts, thereby adding another layer of muddling to individual contributions and the collection as a whole.

Framing

A goal in bringing these muddlings together has been to think about intimacy in feminist geography research while refusing to hoop bands around it to hold particular definitional boundaries in place. Intimacy in feminist geography is not just about social relations or configurations of power, nor is it just about rescaling the mundane as part of reconfiguring globalization. It is also not just about personal writing or the autobiographical. As these muddlings show, like both writings about intimate research acts and intimate research acts themselves, intimacy is a slippery entity that exists among the cracks and folds in the interactions and relationships among people, nonhuman entities and non-living things. In this sense, the writing of intimacy into feminist geography enacted by these intimate research acts are grounded in three affirmative propositions, even though none of them explicitly state so.

First, the methodological approaches in these writings indicate an affinity with a generative ontology that grasps the effects of becoming as a process. Becoming in this sense is not a generic meaning of always in process; rather, becoming is a condition of simultaneously being fixed and fluid, ever-present and fleeting, and durable and transitory. It is productive without being prescriptive. Becoming is not only central to understanding how intimacy works, but also key in the approach to analysis. Whether writing intimately, creatively or more traditionally, the generative constitution of intimacy seeps into the text through the writing itself.

Second, the ontological premise of becoming also applies to the generation of the researcher as a nomadic subject (Braidotti, 2011). As nomads, feminist geography researchers take up numerous and multiple subject positionings, inventing and reinventing themselves as they lurch, stagger and meander through the various practices associated with designing research projects, reading literatures, gathering information and circulating their research in verbal, sensorial and written forms. Some of the contributors detail this process of becoming nomad and bring under critical scrutiny both the role intimacy plays in doing research as well as the effects of intimate acts as part of research. Even so, each contributor has only written about a relatively well-defined project and not about the various positionings that inform an entire research-based academic practice.

Third, part and parcel to becoming nomad is the practice of an affirmative politics. An affirmative politics is about a sustainable transformation individually and collectively. Many of the contributors provide thoughts about individual transformations of a researcher. They do so with a critical eye on how their reflections move along the shared discussion of how to do research. Some of the contributors write about collective transformations that have shaped their research practice.

We see both kinds of contributions as part of a specific academic practice – that is writing intimacy into feminist geography – intended to be a collective endeavour. Even though this endeavour sometimes manifests as individual publications (as it does in this case), it certainly need not always be this way. Although collective, transformative work does not have to be coordinated (for comments on feminist geography, see Moss, 2014; as part of another strand, see Mountz, et al., 2015).

Through their writings, the contributors in this collection show how intimacy is entangled in the affirming research practices of sustainable transformation – emotionally, bodily, collectively and historically – so that personal, social and political change does not destroy that which it desires.

Muddlings

The publication of these chapters in the order presented here – methodological challenges, emergent effects of including one's story, researching intimacy as a topic and analytical methods as part of writing – is but one way to read how these chapters resonate with one another. By way of concluding this volume, we seek to unsettle our own presentation by indicating other types of resonances. In presenting these thoughts on the muddlings, we bear in mind how a feminist geographer might approach research through intimate acts – when designing a project, in the midst of undertaking research or reconsidering on a project that is already completed.

Using intimacy in and through research

Intimacy works through the bodies of researchers and those of their research participants, both humans and nonhumans. Researchers tease out the interconnectedness by exploring specific types of contact or interaction with one's self, another entity or some combination of the two. Kye Askins reflects on how to integrate insights from her own energy practice of *Qigong* and shiatsu into her personal and professional relationships. She shows how intimacy plays out through both her engagement with the people in her life and the nonhuman things she encounters. Her contemplations palpably draw readers into the text as she invites them to breathe with her and respond to both her and their own sensations. Being open to the tone or mood of a situation underlies Kathryn Gillespie's discussion of her research communication with the steers, cows and calves dwelling in her research. Grasping the affect generated by the seemingly routine practices in an animal auction moves her to think through what intimacy has both to say about and to offer research relationships. She works to muddle the notion of what counts as intimacy in order to bring out the complexity of intimacy as an emotion as it plays out in specific circumstances, something she refers to as entangled empathy (after Gruen, 2015). Likewise, Vanessa Massaro and Dana Cuomo track emotional entanglements, not between species but between different groups of people. They reflect on their conflicting emotions over competing intimacies: their relationships, affinities and networks they developed through ethnographic work and their professional communities. They provide examples of how they need to generate ethical as well as practical strategies – sometimes on the spot – to negotiate their ongoing attachments to both groups.

Intimacy in research has the potential to do many things, often at the same time. It can provide a conceptual framing, be a site of inquiry or define a departure point for analysis. The idea of intimacy brings forth an image of specific relations between people and how they might interact and form relationships, even if those are merely intended to be brief encounters, much like a research interview. Intimacy thus can frame an analysis. For example, in her thoughts about some of the interviews she has had with elite global financiers, Maureen Sioh weaves together an account of how postcolonial subject positionings tug at one another, at a psychological level. Although she heard about many things in the interviews that would support her critical analysis of global business, she states that she has chosen not to use them. She is concerned that they might be traced back to an individual. This is more than a matter of confidentiality. She argues that the flip-side of intimacy is moral duplicity, and given the intimate social and cultural connections both she and the men she interviewed occupy, the idea of betraying that which was said in confidence frames her analysis. Another example shows how intimacy can frame how one engages in recording intimate information in the field. Ebru Ustundag calls into question some of the intimacies surrounding self-expression in the common research practice of writing field notes. When there was no clear distinction between the topic of what she was recording in her field journal and the language she used to write it, she began thinking about what field notes were actually *doing*. She comes to understand that field notes as a type of intimate writing constitute her as a research subject, one that is socially mediated through multiple others that take up various positionings in the social structures of power she is trying to dismantle. Oftentimes, intimacy gets taken up more generally as researchers try to disentangle how intimate connections with place inform the context within which the research gets done. Zoë Meletis and Blake Hawkins recount their place-based connections to their research environments and choice of topics. They use a type of personal writing to link their own personal material histories to the choices they made, and continue to make, in their research and career choices as a faculty member and a graduate student.

Intimacy challenges feminist researchers to make sense of how they are involved in the everyday within their own lives and the lives of others. While some of the researchers in this collection stress the inspiring potentiality of intimacy, some write about the restrictive aspects of intimate exposures. Considering the deleterious effects of cultivating intimacy, Courtney Donovan explores the desire to accelerate intimacy through digital health. She queries the pressing of intimacy into both training healthcare professionals and in the doctor–patient relationship and wonders if this may lead to the framing of healthcare choices in terms of personal responsibility. She also raises questions about the desire to impose empathy and intimacy onto professional relationships in the first place. Such an imposition could create a false sense of security by exposing one's own vulnerabilities to healthcare practitioners imbricated within a highly organized knowledge system supported and reproduced by a set of clearly laid-out protocols and practices. The idea of having vulnerabilities exposed through research is often part of the point of doing research. Researchers design projects and inevitably come across those having personal crises. The question becomes an ethical one: what do you do in such situations? In a chance encounter via phone, Gail Adams-Hutcheson and Robyn Longhurst were forced to deal with the threat of suicide. Relocating after an earthquake had driven a woman to despair and she phoned Gail in response to notice about her research project. The extent to which the woman anonymized the exposure of her vulnerability (without providing information about herself or a callback number) prevented Gail from helping someone in danger. Raw exposure to defenceless powerlessness kept Gail as a doctoral student and Robyn as a supervisor from including the phone call in research writing until now.

Intimacy need not necessarily be ensconced in research practices to influence research choices. Exposures of vulnerability arising out of disclosing personal matters in academic settings can shape the way one engages in research as well as a career. In support of a student whose well-being might have been jeopardized if more academic demands were made, Toni Alexander disclosed her own diagnoses of depression and eating disorder to her colleagues. After settling back into the everyday happenings in her department, she was shocked to receive notice that allegations had been made against her about being a security risk: a staff member had used Toni's intimate disclosure of invisible disability against her. The unwarranted claim ignited an extensive formal evaluation process of her research programme. The toll is high for feminist researchers who disclose illness informally and then have that disclosure used against them, such as in invoking formal procedures involving the state apparatus. Yet the state apparatus may be the only institution that can provide justice to those marginalized by a collective indifference, in part because intimacy and the state are entwined at different scales. Sarah de Leeuw draws attention to the murdered and missing Indigenous women in Canada. In advocating for justice, she writes about wider processes of healing for relatives of the women, the communities they lived in and the society that permitted this to happen for so long. The murders and missing person cases of these women have not been solved. The machinations of the local state has thrown them aside simply because they are constituted in society to be disposable.

Productive potential of intimacy

Intimacy pushes one to deal with the edges of everyday encounters, propelling ordinary life events into the extraordinary. These significant, far-reaching encounters can be transformative. Samuel Henkin follows the life of Holocaust survivor Heinz Heger (pen name of Josef Kohout) who tells the story of gay men in Nazi concentration camps. Instead of reducing the stories to a part of history, Samuel wants to hold on to the stories of everyday living in extreme violence, to maintain a living connection, in order to understand the power of postmemory of trauma. Through his intimate writing as part of the process of memorialization, he draws out embodied knowledge that permits him to understand the effects of postmemory, of those events that reach into the future and shape the lives of those who live in the legacies of extreme violence. Life transitions – birth, death, trauma, illness – amplify the transformative potential of embodied knowledge. Clare Madge marks two events, pregnancy and grave illness. In a plea for using artistic expressions as ways to write-up research, she shows how poetry and photography can break open the academic practice of concept development in relation to intimacy. Escaping the habituation of conventional methods of abstraction, she notes that creative bricolage can create spaces that facilitate embodied storytelling, generate insight into emotional intimacies and their relational nature, foster cathartic release through creative agency and create intimate imaginaries in order to release transformative potential.

Intimacy makes articulating transformative potential tricky in practice. In her reflections on crossing cultural norms in the practice of care, Kelsey Hanrahan subtly develops a multidimensional notion of handling for those dying. She comes to understand handling as a mixture of comforting the dying, providing hands-on care and taking on the position of a materialized attendant during death. Kelsey's attending to Nyaa Uchain as she lay dying frees Nyaa Uchain to do things she desires but cannot do because of cultural expectations. Through this process, both Kelsey and Nyaa Uchain become different than they were before the dying began. But sometimes transformation takes a lot more time than a span of a few days or weeks, especially when enveloped in a set of activities that is going faster than the usual everyday pace of life. Pamela Moss writes about negotiating life and death ethics on the fly when confronted with decisions about treatment, empathy, palliative measures and care for her mother. Even though her writing is about what goes on inside the capillaries of the microphysics of biomedical power, she is actually drawn to the commonplaceness of each act even while in the larger picture the collective set of acts career out of anyone's control. She experiences the absurd covalence of a presumed intimacy with her mother's body by physicians alongside her own dislocated intimacy as caregiver and decision-maker for her mother's directives for life and death.

Detailing the effects of intimacy depends on what is being produced through the research. Kathryn Besio, in her sensitive disclosure of a life historically by reading remnants of texts (pictures, written texts) and putting together pieces (momentary snapshots) of a postcolonial puzzle, picks apart what goes into making an intimate gardening space. She likens her research practice of lurking to her critical but sympathetic reading of a family photograph that has been framed by a contemporary politics informed through a complex array of colonial relations. Having followed intimate reverberations across the puzzle pieces, she is able to maintain that there is very little she can say definitively about the intimacies captured in the photograph. While lurking may be one way to pull out and track how intimacies develop historically, there may be other ways to understand how intimacy surfaces within research. Karen Falconer Al-Hindi, Pamela Moss, Leslie Kern and Roberta Hawkins use collective biography to figure out how intimacy emerged in their writing group. They identified events that made them feel closer as a group, more intimate with each other, and then wrote autobiographical pieces capturing their feelings in that moment. They use the analytical notion of inhabiting texts (abandoning meaning in a text to take up what gets created through generating the text) as a way to bring out emergent intimacies. Through their inhabitation of the texts they generated, they were able to distinguish those somewhat intangible things that permitted them to grasp affect, to discern its embodiment and to convey its sustainability as a practice of affirmative politics.

Intimate disclosures, or complex renderings of what counts as intimacy, have consequences. Most remain hidden in contemporary research unless disclosed purposefully as part of a discussion of method and field experience. For Kacy McKinney disclosing her sexuality to her closest field assistants after months of hiding it shows the importance of trust in developing bonds with co-workers. Although she argues that maintaining the relationship in the field was important to her work, there was more at stake than just the immediacy of completing a field season. Women face public pressures in rural India, where they are expected to marry men, have children and raise families. Yet the private spaces, where women can be friends and share secrets about their sexuality and their performances in the field, can be revelatory. For Maral Sotoudehnia, her research site was a set of air-conditioned offices scattered across Dubai while her public everyday life was hidden from view. Her stories of being pulled over by the police and of being mistaken for a Farsi-speaking Persian woman indicate some of the challenges researchers face when living outside their usual living arrangements when doing research. If arrested or abducted, a different version of her story would be circulating for it is not only disclosures about one's identity that matter in the field; other disclosures do too, such as citizenship at border crossings and research topics by asking questions (see Abramson, 2016; see also Bannerjee, 2016).

Closing thoughts

What started out as a short, introductory email from Courtney asking Pamela if she were interested in possibly commenting on a manuscript about graphic memoir turned into a multiple-session conference engagement with personal writing. Through the presentations and engagement with numerous people after the conference, the project then morphed into a proposal for an edited piece of work about intimacy. And now we are at the end of the book, writing the last section, reflecting on the connections that help us figure out what intimacy means, what it does and what it produces. We understand this process as illustrative of the arguments we are making about intimacy: Intimacy is 'bound up in the moments that make us feminist geographers' (Donovan and Moss, this volume, p.3). '[T] through dismantling that which sets up the personal to be personal' (p.10), feminist researchers can figure out what work intimacy has done, what it is doing now and what it can do. Grasping intimacy's potential for change provides a sounder sense of the world and how one might structure interventions to change it.

In offering this collection as part of the intimate turn across geography, we wish to place it within the practice of an affirmative politics. As nomads, feminist geographers can 'set up, sustain and map out sustainable transformations'

(Braidotti, 2011, p.192). These chapters acknowledge and build on an awareness of the role feminist geography has played in bringing about interest in the intimate through engaging the intricacies of the personal as part of everyday life. They show how their reflections on and changes to the research process both inform and generate analyses that open up potential and can bring about change. We have called this project *Writing intimacy into feminist geography*, for writing is – at least for now – a prominent way in which feminist geographers as researchers engage with ideas and with each other. Given the link between intimacy and the personal in both the literature and the everyday, personal writing is widely held as the way to convey the significance of how intimacies work. Yet as these chapters have demonstrated over and over, personal writing is not solely about any one individual. Personal writing accentuates the discursive, affective and material connections people have to other people, nonhuman entities and non-living things. The generative nature of these connections details the process of becoming as part of a relational ontology. Examining intimacy within these connections releases for further scrutiny both the restrictions and the potential the embodied everyday has to inform topics as diverse as communicating research findings to various audiences, interviewing elites and bringing field experience into research write-ups as well as calling for justice for murdered and missing Indigenous women, challenging the practice of biomedicine and cultivating a postmemory of extreme violence. Feminist geographers as researchers take up a panoply of subject positionings to achieve that which they set out to do in their work and their politics. Writing intimacy into feminist geography is part of feminist knowledge production through disciplinary practices both formally, such as designing research and publication, and informally, through communicating with students and colleagues and crafting assignments for courses and workshops. These knowledges and the insights arising from them are part of a collective political project, one that contests oppressive power relations and seeks to transform society.

Being part of this collection has been mesmerizing for us as researchers, academics and colleagues. We have lived in the nooks and crannies of these multifaceted arguments for years now and have still not forgotten the bigger picture of the (decentralizaed and uncoordinated) collective politics of transformation. We hope that you find these works here as captivating as we have.

Works cited

Abramson, Ebby, 2016. Endangered scholars worldwide. *Social Research: An International Quarterly*, 83(1), pp.v–xix.

Bannerjee, Sidhartha, 2016. Canadian academic Homa Hoodfar indicted on unknown charges in Iran. *The Globe and Mail* [online] (Last updated 2.25 PM on 11 July 2016). Available at: http://www.theglobeandmail.com/news/national/canadian-academic-homa-hoodfar-indicted-on-unknown-charges-in-iran/article30849159/ [Accessed 9 August 2016].

Braidotti, Rosi, 2011. *Nomadic theory: the portable Rosi Braidotti*. New York: Columbia University Press.

Braidotti, Rosi, 2013. *The posthuman*. London: Polity Press.

DeVault, Marjorie L., 1999. *Liberating methods: feminism and social research*. Philadelphia: Temple University Press.

Gruen, Lori, 2015. *Entangled empathy*. New York: Lantern Books.

Mol, Annemarie, 2002. *The body multiple: ontology in medical practice*. Durham, NC: Duke University Press.

Moss, Pamela, 2014. Some rhizomatic recollections of a feminist geographer: working toward an affirmative politics. *Gender, Place and Culture*, 21(7), pp.803–12. doi:10.1 080/0966369X.2014.939159

Moss, Pamela and Falconer Al-Hindi, Karen, eds., 2008. *Feminisms in geography: rethinking space, place and knowledges*. Lanham, MD: Rowman and Littlefield.

Mountz, Alison, Bonds, Anne, Mansfield, Becky, Loyd, Jenna, Hyndman, Jennifer, Walton-Roberts, Margaret, Basu, Ranu, Whitson, Risa, Hawkins, Roberta, Hamilton, Trina and Curran, Winnifred, 2015. For slow scholarship: a feminist politics of resistance through collective action in the neoliberal university. *ACME: An International E-Journal for Critical Geographies*, 14(4), [e-journal], pp.1235–59. Available at: http://ojs.unbc.ca/index.php/acme/article/view/1058 [Accessed 11 July 2016].

Index

Note: page numbers in italics refer to illustrations